METALLIC SPINTRONIC DEVICES

Devices, Circuits, and Systems

Series Editor
Krzysztof Iniewski
CMOS Emerging Technologies Research Inc.,
Vancouver, British Columbia, Canada

PUBLISHED TITLES:

Telecommunication Networks
Eugenio Iannone

Testing for Small-Delay Defects in Nanoscale CMOS Integrated Circuits
Sandeep K. Goel and Krishnendu Chakrabarty

Wireless Technologies: Circuits, Systems, and Devices
Krzysztof Iniewski

FORTHCOMING TITLES:

3D Circuit and System Design: Multicore Architecture, Thermal Management, and Reliability
Rohit Sharma and Krzysztof Iniewski

Circuits and Systems for Security and Privacy
Farhana Sheikh and Leonel Sousa

CMOS: Front-End Electronics for Radiation Sensors
Angelo Rivetti

Gallium Nitride (GaN): Physics, Devices, and Technology
Farid Medjdoub and Krzysztof Iniewski

High Frequency Communication and Sensing: Traveling-Wave Techniques
Ahmet Tekin and Ahmed Emira

Labs-on-Chip: Physics, Design and Technology
Eugenio Iannone

Laser-Based Optical Detection of Explosives
Paul M. Pellegrino, Ellen L. Holthoff, and Mikella E. Farrell

Nanoelectronics: Devices, Circuits, and Systems
Nikos Konofaos

Nanomaterials: A Guide to Fabrication and Applications
Gordon Harling, Krzysztof Iniewski, and Sivashankar Krishnamoorthy

Optical Fiber Sensors and Applications
Ginu Rajan and Krzysztof Iniewski

Organic Solar Cells: Materials, Devices, Interfaces, and Modeling
Qiquan Qiao and Krzysztof Iniewski

Power Management Integrated Circuits and Technologies
Mona M. Hella and Patrick Mercier

Radio Frequency Integrated Circuit Design
Sebastian Magierowski

Semiconductor Device Technology: Silicon and Materials
Tomasz Brozek and Krzysztof Iniewski

METALLIC
SPINTRONIC
DEVICES

EDITED BY **XIAOBIN WANG**
Avalanche Technology
Fremont, CA

KRZYSZTOF INIEWSKI MANAGING EDITOR
CMOS Emerging Technologies Research Inc.
Vancouver, British Columbia, Canada

CRC Press
Taylor & Francis Group
Boca Raton London New York

CRC Press is an imprint of the
Taylor & Francis Group, an **informa** business

CRC Press
Taylor & Francis Group
6000 Broken Sound Parkway NW, Suite 300
Boca Raton, FL 33487-2742

© 2015 by Taylor & Francis Group, LLC
CRC Press is an imprint of Taylor & Francis Group, an Informa business

First issued in paperback 2017

No claim to original U.S. Government works
Version Date: 20140425

ISBN 13: 978-1-138-07232-9 (pbk)
ISBN 13: 978-1-4665-8844-8 (hbk)

Library of Congress Cataloging-in-Publication Data

Metallic spintronic devices / edited by Xiaobin Wang.
      pages cm -- (Devices, circuits, and systems ; 32)
    "A CRC title, part of the Taylor & Francis imprint, a member of the Taylor & Francis Group, the academic division of T&F Informa plc."
    Includes bibliographical references and index.
    ISBN 978-1-4665-8844-8 (hardback)
    1. Spintronics--Materials. 2. Magnetic materials. 3. Metals. I. Wang, Xiaobin, 1973-

TK7874.887.M48 2014
621.3--dc23   2014005546

Visit the Taylor & Francis Web site at
http://www.taylorandfrancis.com

and the CRC Press Web site at
http://www.crcpress.com

Contents

Contents

Preface

Semiconductor spintronics are full of fundamental research, and defining their path to unique applications, the metallic spintronic device has been a wealth of important current and emerging applications. Metallic spintronic devices are already in commercial products of computer hard disk drives and memory devices. They hold promise to continue device miniaturization in the postsilicon era and beyond the age of Moore's law.

This book addresses state-of-the-art metallic spintronic device research and development. It covers a wide range of emerging spintronic device applications, from magnetic tunneling junction sensors and spin torque oscillators to spin torque memory and logic. The book is written by top-notch industrial and academic experts, aiming to equip anyone who is serious about metallic spintronics devices with up-to-date design, model, and processing knowledge. The book can be used by either an expert in the field or a graduate student in his or her course curriculum.

The book starts with a review of spintronics applications in current and future magnetic recording devices. Professor Jimmy Zhu is an ABB professor of engineering at Carnegie Mellon University. ABB is a leader in power and automation technologies that enables utility and industry customers to improve performance while lowering environmental impact. The ABB Group of companies operates in close to 100 countries, employing around 120,000 people.

Spin-transfer torque magnetoresistive random access memory (STT-MRAM) device architecture and modeling are presented in Chapter 2 by two industrial experts. Dr. X. Zhu is a staff engineer at Qualcomm Incorporated. Dr. Seung is currently a director of engineering at Advanced Technology, Qualcomm Incorporated. Dr. S. Kang leads STT-MRAM and emerging memory programs for mobile systems. He also serves as distinguished lecturer for the IEEE Electron Device Society. A primary focus of Chapter 2 is on elaborating methodologies to enable robust device and array architectures while addressing challenges in designing variability-tolerant STT-MRAM for advanced systems.

Prospects of STT-MRAM scaling, including detailed multilevel cell structure analysis, are provided in Chapter 3 by a research group at University of Pittsburgh. Professors H. Li and Y. Chen hold more than 100 US patents and have published over 200 technical papers in related areas. Both are the recipients of National Science Foundation (NSF) Career Awards, while Professor Li also received the Young Faculty Award from the US Department of Defense. Y. Zhang and W. Wen are graduate students contributing to this STT-RAM research group.

Memristive effects are universal for spintronic devices at the timescale that explicitly involves the interactions between magnetization dynamics and electronic charge transport. Chapter 4 investigates spintronics device write and read optimization in light of spintronic memristive effects. Dr. X. Wang is a director at Avalanche Technology. He is an advocate and practitioner of top-down and bottom-up approaches to seek memory and data storage solutions to meet the ever-increasing demands on capacity, bandwidth, power, and latency. His research interests include nanoscale magnetoelectronic devices, emerging memory technologies and hybridizing memory/storage architectures, and magnetization dynamics and its implication on spintronics.

In Chapter 5, a group of industrial researchers jointly review the interesting spintronics research directions based upon yttrium iron garnet thin films. These include spin pumping, magnetic proximity, spin hall, and spin Seebeck effects. Dr. Y. Sun currently is with Western Digital, Dr. Z. Wang currently is with Avalanche Technology, and Dr. L. Lu currently is with Seagate Technology. The material in this chapter comes from their joint research work at the Department of Physics, Colorado State University.

Electric field-induced magnetization switching and voltage-controlled magnetic anisotropy effects (Chapter 6) provide unique solutions for low-power spintronics device applications where memory is closely integrated with logic. Professors P. Khalili and L. Wang are well-known experts in this field. Their pioneer work and very early hands-on experiences in this field are widely recognized.

About the Editor

Xiaobin Wang is a director at Avalanche Technology. He received his Ph.D. in the Physics Department and Center for Magnetic Recording Research from the University of California, San Diego. Dr. Wang has worked at Western Digital and Seagate Technology prior to this appointment. He is a consultant at Caraburo Consulting LLC and Ingredients LLC.

Dr. Wang's work in memory and data storage includes device design, advanced technology gap closure, prediction of system performance through bottom-up (from physics to system performance) and top-down (from system performance to component requirements) approaches, company product platform and basic technology roadmap modeling, new concept initiation, and intellectual property analysis. Dr. Wang has published over 100 articles and holds 50 US patents, approved or pending. He has been invited to contribute to various conferences, journals, and books.

Dr. Wang's research interests include nanoscale magnetoelectronic devices, emerging memory technologies and hybridizing memory/storage architectures, and magnetization dynamics and its implication for spintronics. He is an advocate for and practitioner of top-down and bottom-up approaches to seek memory and data storage solutions to meet ever-increasing demands of capacity, bandwidth, power, and latency.

Contributors

Yiran Chen
Department of Electric and
    Computer Engineering
Swanson School of Engineering
University of Pittsburgh
Pittsburgh, Pennsylvania

Seung H. Kang
Qualcomm Technologies
    Incorporated
San Diego, California

Pedram Khalili
Department of Electrical Engineering
University of California, Los Angeles
Los Angeles, California

Hai Li
Department of Electric and
    Computer Engineering
Swanson School of Engineering
University of Pittsburgh
Pittsburgh, Pennsylvania

Lei Lu
Department of Physics
Colorado State University
Fort Collins, Colorado

Yiyan Sun
Department of Physics
Colorado State University
Fort Collins, Colorado

Kang L. Wang
Department of Electrical Engineering
University of California, Los Angeles
Los Angeles, California

Xiaobin Wang
Avalanche Technology
Fremont, California

Zihui Wang
Department of Physics
Colorado State University
Fort Collins, Colorado

Wujie Wen
Department of Electric and
    Computer Engineering
Swanson School of Engineering
University of Pittsburgh
Pittsburgh, Pennsylvania

Yaojun Zhang
Department of Electric and
    Computer Engineering
Swanson School of Engineering
University of Pittsburgh
Pittsburgh, Pennsylvania

Jian-Gang (Jimmy) Zhu
Data Storage Systems Center
Department of Electrical and
    Computer Engineering
Carnegie Mellon University
Pittsburgh, Pennsylvania

Xiaochun Zhu
Qualcomm Technologies
    Incorporated
San Diego, California

1

Perpendicular Spin Torque Oscillator and Microwave-Assisted Magnetic Recording

Jian-Gang (Jimmy) Zhu

Data Storage Systems Center, Department of Electrical and Computer Engineering, Carnegie Mellon University, Pittsburgh, Pennsylvania

CONTENTS

1.1 Introduction

For more than 100 years, virtually all electronic devices had been operated by moving, accumulating, and storing charges carried by either electrons or holes. Spin, the other important property possessed by electrons and holes, besides charge, had largely been ignored in practical applications—until the late 1980s, when the first magnetoresistive sensor based on anisotropic magnetoresistance (AMR) effect was used as a read head in a hard disk drive (HDD) [1]. The magnetic fields from recorded bits in a magnetic thin-film medium was sensed in a thin magnetic film element through a direct resistance change other than measuring magnetic induction that has been used ever since Faraday's 1831 experiment. Accompanied by the excitement from the perspective of commercial technology application, the discovery of the giant magnetoresistance (GMR) effect in 1988 reignited the field of magnetism [2]. The rapid commercialization of the GMR-based read sensor in hard disk drives in the mid-1990s firmly convinced the world that the field of spin-based electronics had begun [3, 4].

One of the fundamental physics concepts behind magnetoresistance effects is the fact that an electron current carries a net spin angular momentum after passing through a magnetized ferromagnetic film. When a spin-polarized current interacts with a local magnetization by exerting a torque on the localized spins, referred to as spin-transfer torque (STT), the orientation of the magnetization could be altered. Following pioneering theoretical work [5, 6], an experimental demonstration in 1999 showed the world for the first time that a spin-polarized current, other than a magnetic field, can be used to switch the magnetization of a ferromagnetic film [7], as many other experiments showed later [8, 9]. Magnetoresistive random access memory (MRAM) based on the STT effect, referred to as STT-MRAM, was quickly proposed along with research and development efforts [10–17]. An MgO-based magnetic tunnel junction has been used as a memory element for both the in-plane magnetization mode and perpendicular magnetization mode [18–20].

It has also been proposed [21] and experimentally investigated that STT can excite spin waves by facilitating spin precession [8, 22–27]. The effect was then quickly utilized to build current-driven magnetic oscillators, with in-plane magnetization mode [24] and perpendicular magnetization mode [25, 28–31], as well as various alternative modes [32, 33].

In this chapter, a design of the perpendicular spin torque oscillator will be reviewed [28]. In particular, its application as a core component for a novel technology, referred to as microwave-assisted magnetic recording, potentially for future hard disk drives, will be discussed in detail [34].

1.2 Perpendicular Spin Torque Oscillator Design Free of External Field

In the perpendicular spin torque oscillator design shown in Figure 1.1, the multilayer stack consists of a spin polarization layer of sufficient perpendicular

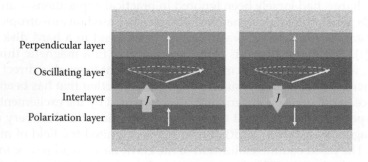

Perpendicular layer

Oscillating layer

Interlayer

Polarization layer

FIGURE 1.1
Schematic drawing of the perpendicular spin torque oscillator design in two oscillating modes, in which the chirality of the magnetization precession in the oscillating layer is solely determined by the magnetization of the perpendicular layer.

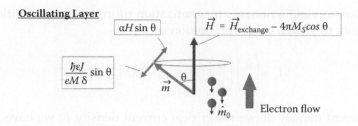

FIGURE 1.2
Spin-transfer torque from spin-polarized current generates an antidamping torque and creates a nonequilibrium condition for the magnetization orientation, yielding a steady gyromagnetization precession of the magnetization in the oscillating layer. (From X. Zhu and J.-G. Zhu, *IEEE Trans. Magn.*, 42(10), 2670–2672, 2006. With permission.)

anisotropy, a nonmagnetic interlayer that is either a metallic or tunnel barrier, and the oscillation layer magnetically coupled to another magnetic layer with perpendicular anisotropy.

Electron current flow through the polarization layer carries net spin angular momentum. The transfer of the angular momentum in the oscillating layer generates a torque on the local magnetization, tilting the magnetization away from the perpendicular direction provided the current density is sufficient. As illustrated in Figure 1.2, at the new balance between the damping torque and the spin-transfer torque, the magnetization no longer is aligned with the local effective magnetic field. The magnetization then naturally precesses around the effective field while maintaining a constant tilting angle at constant current density. The STT-included Landau-Lifshitz-Gilbert equation for describing the gyromagnetic motion of the magnetization in the oscillating layer [6] can be written as

$$\frac{d\hat{m}}{dt} = -\frac{\gamma}{1+\alpha^2}\hat{m}\times\left(\vec{H}-\frac{\alpha\varepsilon\hbar J}{eM\delta}\hat{m}_0\right)-\frac{\gamma}{1+\alpha^2}\hat{m}\times\hat{m}\times\left(\alpha\vec{H}+\frac{\varepsilon\hbar J}{eM\delta}\hat{m}_0\right) \quad (1.1)$$

where the effective magnetic field, *H*, consists of the exchange coupling field arising at the interface with the perpendicular layer and the self-demagnetization field, which depends on the tilting angle θ:

$$H = -\frac{\partial E}{\partial M} = H_{exchange} - 4\pi M_s\cos\theta \quad (1.2)$$

The precessional frequency is then solely determined by the effective field:

$$f = \gamma\cdot H = \gamma\cdot(H_{exchange} - 4\pi M_s\cos\theta) \quad (1.3)$$

Assuming the exchange coupling field is sufficiently greater than the self-demagnetization field, the oscillation will not start until the current density is sufficient to yield a nonzero magnetization tilting angle. The critical current density to yield a steady precession is

$$J_C = \left(\frac{e}{h}\right)\left(\frac{\alpha}{\varepsilon}\right) \cdot (H_{exchange} - 4\pi M_s) \cdot M_s \delta \qquad (1.4)$$

For current density above the critical current density J_c, we have the following linear relationship with current density:

$$f = \gamma \cdot \left(\frac{\hbar}{e}\right)\left(\frac{\varepsilon}{\alpha}\right)\left(\frac{J}{M_s \delta}\right) \qquad (1.5)$$

The above equation shows that the oscillating frequency can be tuned by the current amplitude and linearly increases with increasing current density. This frequency-tunable feature with varying current alone makes this spin torque oscillator extremely interesting for potential technology applications, and the perpendicular spin torque oscillator unique in contrast to that of the in-plane type of spin torque oscillators.

The frequency-tunable range by varying the current is

$$\Delta f = 2\gamma \cdot M_s \qquad (1.6)$$

Keep in mind that the above relation is obtained with the assumption that the magnetization of the perpendicular layer is assumed to be always in the perpendicular direction due to sufficiently strong perpendicular anisotropy in the absence of current. In reality, the frequency-tunable range will be reduced if the perpendicular layer tilts away from the perpendicular direction during the current application.

Micromagnetic modeling studies, which include finite size effect as well as the magnetostatic interactions between all the magnetic layers, have shown some very interesting properties. Figure 1.3 shows simulated STT-driven magnetization precession for a perpendicular spin oscillator design with the perpendicular layer of very high magnetocrystalline anisotropy strength. Increasing current density yields an increase of the precessional frequency due to the accompanying reduction of the self-demagnetizing field as the magnetization becomes more and more in-plane.

The change in the precessional frequency accompanied by the change in the magnetization tilt of the oscillating layer shown in Figure 1.3 clearly demonstrates the mechanism for the frequency tuning ability. Figure 1.4 shows a modeling example for a perpendicular spin torque oscillator (PSTO) design with practical material parameters. In this case, the exchange coupling field is smaller than the self-demagnetization field of the oscillating layer when its magnetization is fully in the perpendicular direction (i.e., when $\theta_1 = 0$), and a nonzero current density will yield magnetization precession of the

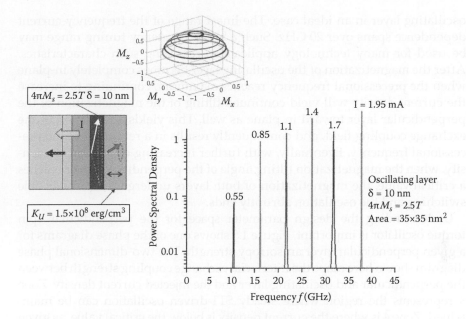

FIGURE 1.3
Modeled STT-driven magnetization precessions for perpendicular spin torque oscillator (PSTO) with a perpendicular layer of strong magnetic anisotropy. The PSTO is designed for generating a high-frequency ac field with a high moment and thick oscillating layer (field-generating layer), intended for the microwave-assisted magnetic recording application. (From J.-G. Zhu et al., *IEEE Trans. Magn.*, 44(1), 125–131, 2008. With permission.)

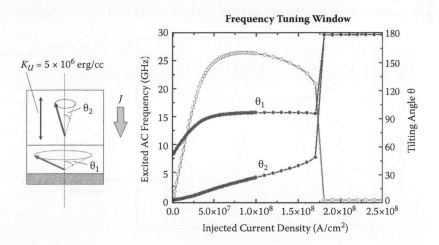

FIGURE 1.4
Calculated current dependence of a perpendicular spin torque oscillator designed with practical material properties. The broad frequency-tunable range is the result of an optimal exchange coupling constant, $A_{ex} = 3.5$ erg/cm^2, assumed in combination with the used anisotropy value in the perpendicular layer. (From X. Zhu and J.-G. Zhu, *IEEE Trans. Magn.*, 42(10), 2670–2672, 2006. With permission.)

oscillating layer in an ideal case. The linear range of the frequency-current dependence spans over 20 GHz. Such a broad frequency tuning range may be used for many technology applications requiring such characteristics. After the magnetization of the oscillating layer becomes completely in-plane when the precessional frequency reaches maximum, continuing to increase the current density will yield continued tilting of the magnetization of the perpendicular layer toward in-plane as well. This yields a reduction of the exchange coupling field, and consequently results in a reduction of the precessional frequency. Eventually, with further increasing of the current density, when the magnetization tilting angle of the perpendicular layer reaches a critical value, the magnetization of both layers undergoes an irreversible switching and the oscillation abruptly ends.

Understanding the design parameter space for the perpendicular spin torque oscillator is important. Figure 1.5 shows one of the phase diagrams for a given perpendicular layer anisotropy strength. The two-dimensional phase diagram shown here maps the interfacial exchange coupling strength between the perpendicular and oscillating layer and the injected current density. Zone 3 represents the region where steady STT-driven oscillation can be maintained. Zone 4 is where the current density is below the critical value, as given by Equation (1.4). Zone 5 indicates the region where the current density has exceeded the critical value where the magnetizations of both layers have irreversibly switched to reach an equilibrium state where no oscillation will occur. In zone 1, the oscillating layer is irreversibly switched by the current density being high relative to the exchange coupling. Although zone 2 is labeled with steady precession in this figure, a later study showed that the configuration

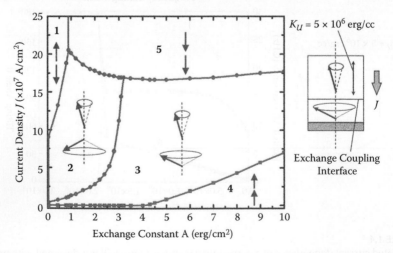

FIGURE 1.5
Calculated phase diagram for the exchange coupling strength and current density. Zone 3 represents the normal oscillation region. (From X. Zhu and J.-G. Zhu, *IEEE Trans. Magn.*, 42(10), 2670–2672, 2006. With permission.)

for elements with not-so-small size yields both spatial and temporal chaotic magnetization precession with a significantly broadened oscillation spectrum.

All the predicated features shown above have essentially been demonstrated by various experimental investigations on the proposed perpendicular spin torque oscillator. Figure 1.6 shows the measured power spectral density at a series of injected current values for a perpendicular spin torque oscillator. The oscillating layer is a 3 nm thick Co layer that is exchange coupled with a Co/Pd multilayer with an anisotropy field of 5 kOe. The exchange coupling is significant such that the Co oscillating layer is perpendicular in the absence of the applied field. The oscillator is patterned into a 50 nm diameter pillar using e-beam-based lithography with ion-million. A critical current of 5 mA is needed to start the oscillation at 5.1 GHz, and beyond that, the oscillating frequency increases with increasing the current until a maximum frequency of 6.2 GHz. The $\Delta f = 1.1$ GHz current-tunable frequency range is less than half of what is predicted by Equation (1.4), indicating significant magnetization tilting of the perpendicular layer at these current values, especially at 8.5 mA, where the oscillation frequency starts to decrease as current continues to increase. Note that the linewidth of the spectrum at 8 mA injecting current is extremely narrow, reflecting a characteristic feature of the steady STT-driven oscillation. In contrast to what has been reported in Ref [24], where the perpendicular layer was absent, the linewidth shown in Figure 1.6 is significantly narrower with the perpendicular layer and sufficient exchange coupling.

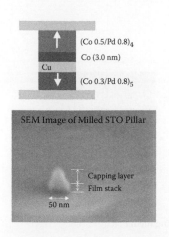 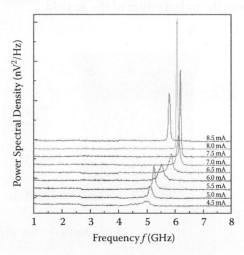

FIGURE 1.6
Left: Schematic drawing and SEM image of a fabricated perpendicular spin torque oscillator nanopillar. Right: Measured power spectral densities of the current-driven magnetization oscillation from one of the fabricated spin torque oscillators. A perpendicular field of 600 Oe is applied in the direction of the magnetization of the perpendicular layer to compensate the coupling field arising from the bottom polarization layer. (From C. H. Sim et al., *J. Appl. Phys.*, 111(7), 07C914, 2012. With permission.)

1.3 Microwave-Assisted Magnetic Recording

The fact that magnetization in the oscillating layer can precess in the film plane makes the perpendicular oscillator ideally suited for generating a localized ac magnetic field at the microwave frequency regime [34]. This unique capability can be used in magnetic recording as a future technology for hard disk drives [34–36]. Figure 1.7 shows a schematic drawing of a recording head with a perpendicular spin torque oscillator embedded into the training shield gap. The generated ac magnetic field in the medium would provide substantial assistance to the recording head field, enabling high-fidelity recording in the medium with effective anisotropy substantially greater than that in the recording head field.

To generate an ac field of significant amplitude inside recording media, the oscillating layer, also referred to as the field-generating layer (FGL), needs to be relatively thick, around 10 to 15 nm, and have large saturation magnetization, greater than 2 T. If the technology is applied to the next generation of head disk drives, the intended track width will be around 40 nm. With such a narrow track width, the magnetic poles on the side edges of the FGL when the magnetization of the FGL is horizontal also produce a significant field in the media. Consequently, the generated ac field on the recording track in the media becomes circularly polarized, as illustrated in Figure 1.8. The circles in the figure indicate the trajectory of the STO-generated ac field vectors— hence the polarization of the ac field.

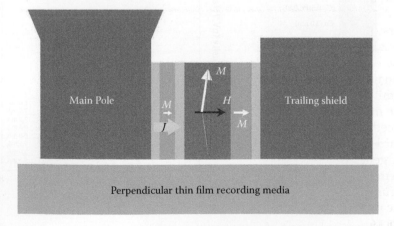

FIGURE 1.7
Schematic drawing of a perpendicular STO embedded in the write-gap of a recording head. The magnetization precession in the oscillating layer will generate a high-frequency ac field in the thin-film recording media. The stray field strength inside the write-gap is around 1.2 T. If the required ac field frequency is below 33 GHz, the perpendicular layer in the STO can be omitted.

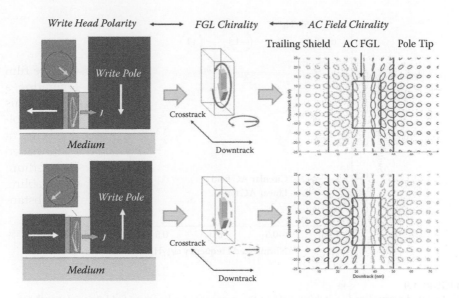

FIGURE 1.8
Illustration of the circular ac field generated by the oscillating layer of the PSTO in the write-gap. The circles in the pictures on the right indicate the field vector trajectory mapped in the plane of the recording medium, with the different color differentiating opposite chirality of the rotating field.

By applying the ac field, saturation recording can be achieved on a medium with significantly higher anisotropy with the same recording head field. The implication could be very significant. Over the past two decades, the area data storage density in a hard disk drive has increased by almost three orders of magnitude, from 1 Gbit/in.² to nearly 1 Tbits/in.². This advancement is mainly achieved by increasing the switching field of magnetic grains in the recording media so that grain size can be reduced without suffering thermal erasure. However, the recording head field magnitude has reached its limitation, and we will no longer be able to increase the switching field of the medium grains. As a result, the medium grain size cannot be further reduced without weakening thermal stability, and the area density growth in hard disk drives has slowed to less than 15% per year for 2012. There is an urgent need for a technology breakthrough to renew the area density growth we have enjoyed in the past.

Figure 1.9 gives a quantitative description of the switching field reduction in the presence of either linear or circular ac fields. Here, the recording field is a pulsed field with 1 ns pulse duration and 0.1 ns rise time, and is applied 30° off the easy axis of the grain, opposite the initial magnetization. The calculated switching field is normalized to the anisotropy field of the grain, H_k, and the ac field is continuous and its frequency is normalized to γH_k.

As shown in the figure, the circular ac field is much more effective in terms of switching field reduction with respect to grain anisotropy field than that

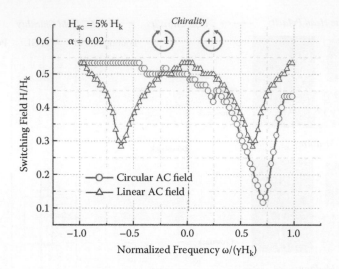

FIGURE 1.9
Calculated normalized switching field in the presence of a linear ac field (triangle) and a circular ac field (circle). The magnitude of the ac field is kept at 1/20 of the grain anisotropy field.

in the presence of the linear ac field. However, for the circular ac field, the chirality has to match the chirality of the grain magnetization precession prior to switching. It turns out that the chirality of the ac field generated by the PSTO embedded (Figure 1.8) in the write-gap matches the required chirality correctly for assisting the recording process. It is also important to note that the frequency range where the switching field reduction is significant is relatively broad [34, 37].

The effectiveness of a circular ac field over that of a linear ac field can be viewed as the following. A linear ac field can be decomposed as the superposition of two circular ac field components with opposite chirality [38, 39]:

$$H_0 \cos \omega t \cdot \hat{e}_x = H_0 \left(\frac{1}{2} (\cos \omega t \cdot \hat{e}_x + \sin \omega t \cdot \hat{e}_y) + \frac{1}{2} (\cos \omega t \cdot \hat{e}_x - \sin \omega t \cdot \hat{e}_y) \right) \quad (1.7)$$

Because the component with the chirality opposite to the chirality of the grain magnetization precession (prior to switching) has zero effect on switching, only the component with chirality matching the precession is effective, and its magnitude is only half.

Examining the magnetization trajectory of a grain during the ac field-assisted switching also helps to illustrate the effectiveness of a circular ac field over a linear one. Figure 1.10 shows two representative cases for switching under a circular ac field and a linear ac field of the same magnitude. During switching, the point at which the magnetization has become irreversibly switched can be identified as the switching point where the chirality of the grain magnetization precession reverses, as it can be observed in

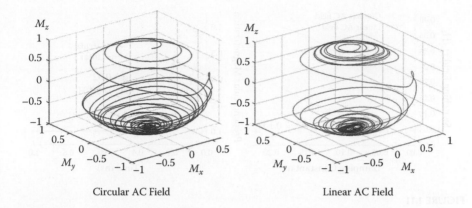

Circular AC Field   Linear AC Field

FIGURE 1.10
Simulated magnetization trajectory of a grain during magnetization switching in the presence of a circular ac field (left) and a linear ac field. The magnitude of the two ac fields is the same. The irreversible points during switching, where the effective magnetic field of the grain reverses its direction, occur when the chirality of the magnetization precession reverses.

each case in the figure. The number of cycles of the magnetization precession prior to the switching point is significantly less for the circular ac field case than for the linear ac field case.

The ac field-assisted switching can be characterized as switching below coercivity. Although the recording field is insufficient to remove the energy barrier (as it does in the case of conventional recording), the presence of a circular ac field component with chirality matching that of the grain magnetization precession enables the grain magnetization to gain energy to overcome the energy barrier, as in the example shown in Figure 1.12. Thus, the intrinsic damping of the grain magnetic energy is important because it describes the rate of energy "leaking" during the assisted switching process. As shown in the left graph of Figure 1.11, increasing the Gilbert damping constant will make the ac field assist less effective with smaller switching field reduction. However, the rise of the switching field in the circular ac field case is much smaller than that in the linear ac field case.

More in-depth understanding of the effect of the Gilbert damping constant is important. Figure 1.13 shows calculated switching field as a function of ac field frequency for a segmented grain with two exchange break layers (EBLs) [40–43]. In the figure, three cases with three different values of the Gilbert damping constant for the grain are shown. Again, the recording field is a 1 ns pulse, applied in the direction 30° off the anisotropy easy axis, while a circular ac field is applied continuously. The red cell in the map indicates a magnetization reversal, and blue otherwise. The switching field of the grain without the ac field is around 12,000 Oe. Note that a higher value of the damping constant mainly yields the shift of the boundary region of the switching map toward lower frequencies, while the switching map at lower frequencies is virtually unchanged. The implication of these results is

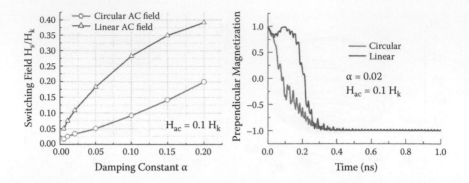

FIGURE 1.11
Left: Calculated switching field in the presence of a circular ac field (circles) and a linear ac field (triangles) as a function of the Gilbert damping constant of the grain. Right: Transient process of the easy axis magnetization component during switching for the case with the circular ac field (thinner) and the linear ac field (thicker). The recording field is a 1 ns pulse at the direction 30° off the easy axis. (From J. Zhu and Y. Wang, *IEEE Trans. Magn.*, 46(3), 751–757, 2010.)

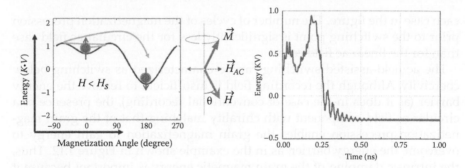

FIGURE 1.12
Illustration of the mechanism for switching field reduction in the presence of an ac field. Left: Schematic drawing of grain magnetic energy as a function of magnetization angle in the presence of a recording field H. Right: The magnetic energy of a grain during switching in the presence of a linear ac field.

the following: to operate at not-so-high ac field frequencies could eliminate or significantly alleviate the impact of damping constant dispersion among the grains in media. In practice, one should avoid operating near the high ac field frequency boundary region of the switching map, where the switching is marginal and sensitive to intrinsic parameter variations. Recording processes on perpendicular granular thin-film media with segmented grains have been simulated by superimposing a practical recording head field and the ac field generated by the PSTO embedded in the write-gap of the recording head. Figure 1.14 shows an example of the simulation, in which case each grain has three 4-nm-thick segments in the perpendicular (along grain height) direction with optimized exchange coupling between adjacent segments. It is interesting to note that the recording footprint is no longer the

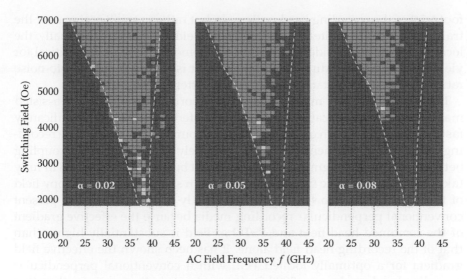

FIGURE 1.13
See color insert. Calculated switching maps for a segmented grain with two exchange break layers inserted. The reversal field is a 1 ns pulse at the direction 30° off the grain anisotropy easy axis, while a steady circular ac field is applied in the horizontal plane. A red cell indicates irreversible magnetization switching, and blue otherwise. The switching field of the grain without the ac field is 12,000 Oe.

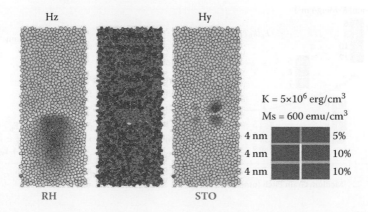

FIGURE 1.14
See color insert. Map of the perpendicular (left) and down-track (right) components of the recording field along with the resulting transition pattern (middle). The head moves down relative to the medium. x is in the cross-track direction, y the down-track direction, and z the perpendicular direction.

footprint of the recording head write-pole, but rather a small region near the trailing edge of the write-pole where the ac field is significant. Actually, the location of the FGL inside the write-gap becomes important and critical for yielding optimal recording performance, that is, an optimal signal-to-noise ratio (SNR) of the resultant recording data pattern.

The segmentation of medium grain is important for microwave-assisted magnetic recording. Because the PSTO-generated ac field decays significantly faster through the depth of the media away from the surface than the recording head field, the segmentation with adequately reduced exchange coupling between the surface segment and the segment below will enable one to fully take advantage of the ac field near the medium surface. The anisotropy field of the top segment also should be significantly higher than that of present conventional perpendicular recording media because the effective gradient of the combined head field and PSTO ac field is substantially higher than that of the recording head field alone. It has been shown the effective field gradient for a optimally located FGL with a conventional perpendicular recording head can be over 3000 Oe/nm, at least six times greater than that of an optimized conventional recording head without PSTO. This factor is of great importance because a significantly enhanced write field gradient is also essential for enabling areal density capability (ADC) gain in addition to the writability gain [39].

To reduce grain size, the anisotropy energy constant of the grain needs to be sufficiently high to ensure sufficient thermal stability. Figure 1.15 shows segmented grain stack designs for different grain pitches, and thereby different grain sizes. In each case, the anisotropy strength of the segments is chosen such that the energy barrier, E_b, of an isolated grain always exceeds

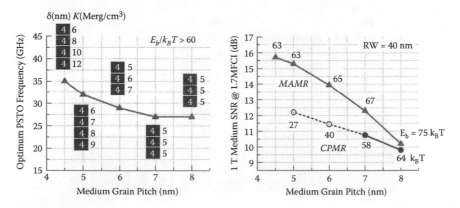

FIGURE 1.15
Left: Segmented grain stack designs for different grain sizes. The grain size is assumed to be 95% of the grain pitch. The vertical exchange coupling through each EBL is optimized for obtaining maximum SNR. Right: SNR at 1.7 MFCI as a function of mean grain pitch for the designs shown on the left. The read track width is kept at 40 nm, and the written track width is approximately 50 nm.

60 $k_B T$. The actual energy barrier should be higher at film level due to the small intergranular exchange coupling among the top segments of the adjacent grains. The through EBL exchange coupling is optimized between the segments for each grain stack design to obtain maximum SNR. The insertion of EBL does yield a slight reduction in the energy barrier compared to simply adding energy barriers of the segments within a grain [44]. However, this reduction becomes smaller for higher-anisotropy designs at smaller grain sizes because the vertical coupling through EBL needs to be stronger.

Note that the anisotropy energy constant of the top segment is 6×10^6 erg/cm^3 or lower because recording becomes difficult even with the assistance of the STO-generated ac field. The segmentation by inserting the EBLs enables grain switching of lower segments with higher anisotropy strength, similar to today's medium design, to maintain the grain energy barrier as grain size reduces. With the segmented design, increasing the energy barrier does not require an equal amount of writability enhancement. As a result, the optimum ac field frequency only increases slightly as grain size decreases. Only when grain pitches become smaller than 5 nm does the STO design need to have a perpendicular anisotropy layer exchange coupled to the FGL to produce an oscillation frequency higher than that supported by the in-gap field. Otherwise, the perpendicular anisotropy layer can be omitted.

With the ac field produced by the PSTO at the corresponding frequencies, excellent recording can be obtained for all the medium designs shown in the figure. The right graph shows the calculated SNR for all the designs shown in the left graph. The SNR performance of the optimized perpendicular magnetic recording (PMR) medium grain stack design is also plotted in the figure, even though its energy barrier is too low to be viable at small grain pitches. The numbers shown at each data point indicate the corresponding energy barrier in units of $k_B T$, with T = 300 K. Again, these energy barrier values are for individual grains. In the case of MAMR, SNR increases with decreasing grain pitch, and at small grain pitches, the SNR is substantially higher than that of the PMR case. At grain pitches of 8 nm and beyond, present PMR essentially creates grain pitch-limited transitions and MAMR as well. At smaller grain pitches, transitions created in the MAMR case become sharper than those in the PMR case, yielding higher signal and lower jitter noise due to the significantly higher effective write field gradient. It is important to note that enhanced writability alone is not sufficient for ADC gain, which relies on the ability to produce grain pitch-limited transitions at small grain pitches. Both modeling studies and experimental investigation have shown MAMR at optimal conditions will significantly enhance the effective write field gradient, and hence enables creation of grain pitch-limited transitions even at very small grain pitches [35, 36, 39, 45].

Acknowledgment

The author thanks Drs. Xiaochun Zhu, Yuhui Tang, Yiming Wang, Cheow Hin Sim, and Matt Moneck for their contribution to this work while at Carnegie Mellon University. The research has been supported in part by the Data Storage Systems Center at Carnegie Mellon and some of its industrial sponsors, and by the National Science Foundation MRSEC through Johns Hopkins University.

References

1. T. McGuire and R. Potter, Anisotropic magnetoresistance in ferromagnetic 3d alloys, *IEEE Trans. Magn.*, 11(4), 1018–1038, 1975.
2. M. N. Baibich, J. M. Broto, A. Fert, F. Nguyen Van Dau, F. Petroff, P. Eitenne, G. Creuzet, A. Frierich, and J. Chazelas, Giant magnetoresistance of (001)Fe/(001) Cr magnetic superlattice, *Phys. Rev. Lett.*, 61(1), 2472–2475, 1988.
3. B. Dieny, V. S. Speriosu, B. A. Gurney, S. S. P. Parkin, D. R. Wilhoit, K. P. Roche, S. Metin, D. T. Peterson, and S. Nadimi, Spin-valve effect in soft ferromagnetic sandwiches, *J. Magn. Magn. Mater.*, 93, 101–104, 1991.
4. C. Tsang, R. E. Fontana, T. Lin, D. E. Heim, V. S. Speriosu, B. A. Gurney, and M. L. Williams, Design, fabrication and testing of spin-valve read heads for high density recording, *IEEE Trans. Magn.*, 30(6), 3801–3806, 1994.
5. L. Berger, Emission of spin waves by a magnetic multilayer traversed by a current, *Phys. Rev. B*, 54(13), 9353–9358, 1996.
6. J. C. Slonczewski, Current-driven excitation of magnetic multilayers, *J. Magn. Magn. Mater.*, 159(1–2), L1–L7, 1996.
7. E. B. Myers, Current-induced switching of domains in magnetic multilayer devices, *Science*, 285(5429), 867–870, 1999.
8. J. Katine, F. Albert, R. Buhrman, E. Myers, and D. Ralph, Current-driven magnetization reversal and spin-wave excitations in Co/Cu/Co pillars, *Phys. Rev. Lett.*, 84(14), 3149–3152, 2000.
9. J. Grollier, V. Cros, A. Hamzic, J. M. George, H. Jaffrès, A. Fert, G. Faini, J. Ben Youssef, and H. Legall, Spin-polarized current induced switching in Co/Cu/Co pillars, *Appl. Phys. Lett.*, 78(23), 3663, 2001.
10. M. Hosomi, H. Yamagishi, T. Yamamoto, K. Bessho, Y. Higo, K. Yamane, H. Yamada, M. Shoji, H. Hachino, C. Fukumoto, H. Nagao, and H. Kano, A novel nonvolatile memory with spin torque transfer magnetization switching, *IEDM Tech. Dig.*, 459–462, 2005.
11. X. Zhu and J.-G. Zhu, Spin torque and field-driven perpendicular MRAM designs scalable to multi-Gb/chip capacity, *IEEE Trans. Magn.*, 42(10), 2739–2741, 2006.

12. K. Miura, T. Kawahara, R. Takemura, J. Hayakawa, S. Ikeda, R. Sasaki, H. Takahashi, H. Matsuoka, and H. Ohno, A novel SPRAM (SPin-transfer torque RAM) with a synthetic ferrimagnetic free layer for higher immunity to read disturbance and reducing write-current dispersion, in *Symposium on VLSI Technology Digest of Technical Papers*, June 2007, pp. 234–235.

13. T. Kawahara, R. Takemura, K. Miura, J. Hayakawa, S. Ikeda, Y. Lee, R. Sasaki, Y. Goto, K. Ito, T. Meguro, F. Matsukura, H. Takahashi, H. Matsuoka, H. Ohno, 2Mb Spin-transfer torque RAM (SPRAM) with bit-by-bit bidirectional current write and parallelizing-direction current read, *ISSCC* 2007, Paper 26.5, 2007.

14. R. Beach, T. Min, C. Horng, Q. Chen, P. Sherman, S. Le, K. Yang, H. Yu, X. Lu, W. Kula, T. Zhong, R. Xiao, A. Zhong, G. Liu, J. Ken, J. Yuan, J. Chen, R. Tong, J. Chien, T. Torng, P. Wang, M. Chen, S. Assefa, M. Qazi, J. Debrosse, M. Gaidis, S. Kanakasabapathy, Y. Lu, J. Nowak, E. O. Sullivan, T. Maffitt, J. Z. Sun, and W. J. Gallagher, A statistical study of magnetic tunnel junctions for high-density spin torque transfer-MRAM (STT-MRAM), in *IEDM 2008*, December 2008, pp. 1–4.

15. C. J. Lin, S. H. Kang, Y. J. Wang, K. Lee, X. Zhu, W. C. Chen, X. Li, W. N. Hsu, Y. C. Kao, M. T. Liu, M. Nowak, and N. Yu, 45nm low power CMOS logic compatible embedded STT MRAM utilizing a reverse-connection 1T/1MTJ cell, in *2009 IEEE International Electronic Devices Meeting*, December 2009, pp. 1–4.

16. K. Tsuchida, T. Inaba, K. Fujita, Y. Ueda, T. Shimizu, Y. Asao, T. Kajiyama, M. Iwayama, K. Sugiura, S. Ikegawa, T. Kishi, T. Kai, M. Amano, N. Shimomura, H. Yoda, and Y. Watanabe, 14.2 A 64Mb MRAM with clamped-reference and adequate-reference schemes, in *2010 IEEE International Solid-State Circuits Conference*, 2010, pp. 138–139.

17. D. C. Worledge, G. Hu, P. L. Trouilloud, D. W. Abraham, S. Brown, M. C. Gaidis, J. Nowak, E. J. O'Sullivan, R. P. Robertazzi, J. Z. Sun, and W. J. Gallagher, Switching distributions and write reliability of perpendicular spin torque MRAM, in *2010 International Electronic Devices Meeting*, December 2010, pp. 12.5.1–12.5.4.

18. S. Yuasa, T. Nagahama, A. Fukushima, Y. Suzuki, and K. Ando, Giant room-temperature magnetoresistance in single-crystal Fe/MgO/Fe magnetic tunnel junctions, *Nat. Mater.*, 3(12), 868–871, 2004.

19. S. S. P. Parkin, C. Kaiser, A. Panchula, P. M. Rice, B. Hughes, M. Samant, and S.-H. Yang, Giant tunnelling magnetoresistance at room temperature with MgO (100) tunnel barriers, *Nat. Mater.*, 3(12), 862–867, 2004.

20. Y. Huai, F. Albert, P. Nguyen, M. Pakala, and T. Valet, Observation of spin-transfer switching in deep submicron-sized and low-resistance magnetic tunnel junctions, *Appl. Phys. Lett.*, 84(16), 3118, 2004.

21. L. Berger, Emission of spin waves by a magnetic multilayer traversed by a current, *Physical Review B.*, 54, 9353–9358, 1996.

22. M. Tsoi, A. Jansen, J. Bass, W.-C. Chiang, M. Seck, V. Tsoi, and P. Wyder, Excitation of a magnetic multilayer by an electric current, *Phys. Rev. Lett.*, 80(19), 4281–4284, 1998.

23. M. Tsoi, A.G.M. Jansen, J. Bass, W. Chiang, V. Tsoi, and P. Wyder, Generation and detection of phase-coherent current-driven magnons in magnetic multilayers, *Nature*, 406, 46–48, 2000.

24. S. I. Kiselev, J. C. Sankey, I. N. Krivorotov, N. C. Emley, R. J. Schoelkopf, R. A. Buhrman, and D. C. Ralph, Microwave oscillations of a nanomagnet driven by a spin-polarized current, *Nature*, 425(6956), 380–383, 2003.

25. S. Kiselev, J. Sankey, I. Krivorotov, N. Emley, M. Rinkoski, C. Perez, R. Buhrman, and D. Ralph, Current-induced nanomagnet dynamics for magnetic fields perpendicular to the sample plane, *Phys. Rev. Lett.*, 93(3), 036601, 2004.
26. W. Rippard, M. Pufall, S. Kaka, S. Russek, and T. Silva, Direct-current induced dynamics in Co90Fe10/Ni80Fe20 point contacts, *Phys. Rev. Lett.*, 92(2), 027201, 2004.
27. W. Rippard, M. Pufall, S. Kaka, T. Silva, and S. Russek, Current-driven microwave dynamics in magnetic point contacts as a function of applied field angle, *Phys. Rev. B*, 70(10), 100406, 2004.
28. X. Zhu and J.-G. Zhu, Bias-field-free microwave oscillator driven by perpendicularly polarized spin current, *IEEE Trans. Magn.*, 42(10), 2670–2672, 2006.
29. D. Houssameddine, U. Ebels, B. Delaët, B. Rodmacq, I. Firastrau, F. Ponthenier, M. Brunet, C. Thirion, J.-P. Michel, L. Prejbeanu-Buda, M.-C. Cyrille, O. Redon, and B. Dieny, Spin-torque oscillator using a perpendicular polarizer and a planar free layer, *Nat. Mater.*, 6(6), 441–447, 2007.
30. C. H. Sim, S. Y. H. Lua, T. Liew, and J.-G. Zhu, Current driven oscillation and switching in Co/Pd perpendicular giant magnetoresistance multilayer, *J. Appl. Phys.*, 109(7), 07C905, 2011.
31. C. H. Sim, M. Moneck, T. Liew, and J.-G. Zhu, Frequency-tunable perpendicular spin torque oscillator, *J. Appl. Phys.*, 111(7), 07C914, 2012.
32. V. Pribiag, G. Finocchio, B. Williams, D. Ralph, and R. Buhrman, Long-timescale fluctuations in zero-field magnetic vortex oscillations driven by dc spin-polarized current, *Phys. Rev. B*, 80, 180411, 2009.
33. S. Kaka, M. R. Pufall, W. H. Rippard, T. J. Silva, S. E. Russek, and J. A. Katine, Mutual phase-locking of microwave spin torque nano-oscillators, *Nature*, 437(7057), 389–392, 2005.
34. J.-G. Zhu, X. Zhu, and Y. Tang, Microwave assisted magnetic recording, *IEEE Trans. Magn.*, 44(1), 125–131, 2008.
35. M. Matsubara, M. Shiimoto, K. Nagasaka, Y. Sato, Y. Udo, K. Sugiura, M. Hattori, M. Igarashi, Y. Nishida, H. Hoshiya, K. Nakamoto, and I. Tagawa, Experimental feasibility of spin-torque oscillator with synthetic field generation layer for microwave assisted magnetic recording, *J. Appl. Phys.*, 109(7), 07B741, 2011.
36. Y. Nozaki, N. Ishida, Y. Soeno, and K. Sekiguchi, Room temperature microwave-assisted recording on 500-Gbpsi-class perpendicular medium, *J. Appl. Phys.*, 112(8), 083912, 2012.
37. Y. Wang, Y. Tang, and J.-G. Zhu, Media damping constant and performance characteristics in microwave assisted magnetic recording with circular ac field, *J. Appl. Phys.*, 105(7), 07B902, 2009.
38. M. Igarashi, Y. Suzuki, H. Miyamoto, Y. Maruyama, and Y. Shiraishi, Effect of elliptical highfrequency field on microwave-assisted magnetic switching, *IEEE Trans. Magn.*, 45(10), 3711–3713, 2009.
39. J. Zhu and Y. Wang, Microwave assited magnetic recording utlizing perpendicular spin torque oscillator with switchable perpendicular electrodes, *IEEE Trans. Magn.*, 46(3), 751–757, 2010.
40. R. H. Victora and X. Shen, Exchange coupled composite media for perpendicular magnetic recording, *IEEE Trans. Magn.*, 41(10), 2828–2833, 2005.
41. J. Wang, W. Shen, and J. Bai, Exchange couple composite media for perpendicular magnetic recording, *IEEE Trans. Magn.*, 41(10), 3181–3186, 2005.

42. K. Tang, X. Bian, G. Choe, K. Takano, M. Mirzamaani, G. Wang, J. Zhang, Q.-F. Xiao, Y. Ikeda, J. Risner-Jamtgaard, and X. Xu, Design consideration and practical solution of high-performance perpendicular magnetic recording media, *IEEE Trans. Magn.*, 45(2), 786–792, 2009.

43. D. Suess, T. Schrefl, S. Fähler, M. Kirschner, G. Hrkac, F. Dorfbauer, and J. Fidler, Exchange spring media for perpendicular recording, *Appl. Phys. Lett.*, 87(1), 012504, 2005.

44. J. J. Zhu and Y. Wang, SNR enhancement in segmented perpendicular media, *IEEE Trans. Magn.*, 47(10), 4066–4072, 2011.

45. M. Shiimoto, M. Igarashi, M. Sugiyama, Y. Nishida, and I. Tagawa, Effect of effective field distribution on recording performance in microwave assisted magnetic recording, *IEEE Trans. Magn.*, 49(7), 3636–3639, 2013.

2

Spin-Transfer-Torque MRAM:
Device Architecture and Modeling

Xiaochun Zhu and Seung H. Kang

Qualcomm Technologies Incorporated, San Diego, California

CONTENTS

2.1 Introduction

Conventional electronic devices made of silicon rely on the transport of electrons as charge carriers. Electrons have another important property, spin, which makes each electron behave like a minuscule magnet. Spintronics is a field of science and engineering involving both attributes of electrons in an attempt to build a useful device [1]. The spintronic phenomenon that triggered extensive R&D efforts is known as giant magnetoresistance (GMR), which was first reported in 1988 by two European research groups independently: one led by Albert Fert in France [2] and the other by Peter Grünberg in Germany [3]. The discovery of GMR promptly opened an emerging field of technology, spintronics, on magnetoelectric multilayers and devices. In 2007, the Nobel Committee awarded both scientists the Nobel Prize for Physics.

In recent years, technologists engineered more advanced magnetoelectric multilayers, magnetic tunnel junctions (MTJs), by integrating a tunnel barrier in lieu of a metallic nonmagnetic spacer. Tunnel magnetoresistance [4] is now a well-understood quantum mechanical phenomenon that leads to a hysteresis behavior of MTJ, as illustrated in Figure 2.1.

Until now, the most significant application of spintronics has been accomplished by the mass storage device industry, primarily in hard disk drives (HDDs). Advances in perpendicular magnetic anisotropy (PMA) recording media and read head sensors have enabled the use of HDDs with ever-increasing data storage capacity on the order of terabytes at ever-lower cost. However, HDDs require mechanical components such as spinning disks and moving read and write heads, causing drawbacks such as limited access time, large form factor, and susceptibility to mechanical reliability problems. This opened a path for a solid-state drive, currently based on NAND Flash

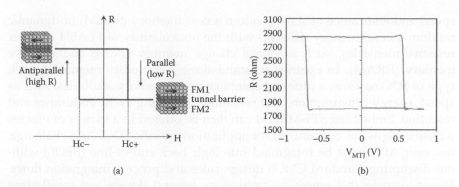

FIGURE 2.1
(a) Conceptual illustration of a hysteresis curve showing the MTJ resistance (R) as a function of the magnetic field (H). A binary resistance state can be obtained through making the magnetization of the upper ferromagnetic layer (FM1, free layer) with respect to that of the lower ferromagnetic layer (FM2, pinned or reference layer) either parallel or antiparallel at the switching field H_{c+} or H_{c-}, respectively. (b) An experimental observation of an MTJ hysteresis curve as a function of the electric voltage (V) for which switching is driven by current (STT).

memory, to enter the data storage market, in particular for consumer products such as smartphones, tablets, and notebook computers.

Recently, technologists started developing more advanced spintronic devices built on a semiconductor integrated circuit (IC) platform. A widely publicized device is magnetoresistive random access memory (MRAM). The first MRAM product was commercialized in 2006 by EverSpin Technologies (then as part of Freescale Semiconductor) [5]. While the memory capacity was small (4 Mbits) and the IC was based on a legacy 180 nm complementary metal oxide semiconductor (CMOS) platform, the enabling technology (Toggle MRAM [6, 7]) was regarded as a remarkable by-product of fundamental nanomagnetism, sophisticated MTJ engineering, and tailored circuit design. Until now, however, the commercially available Toggle MRAM has not made the impact originally anticipated. Drawbacks pertaining to this first-generation MRAM are well understood, limiting its application to a niche market.

The spintronics community has achieved significant scientific discoveries and engineering breakthroughs to resolve such drawbacks. Most recognized is the emergence of spin-transfer-torque (STT)-MRAM. Key findings and advances have triggered industry-wide R&D efforts in pursuit of an alternative memory in lieu of conventional memories that are not only facing acute trade-offs in performance and power, but also nearing fundamental scaling limits. In parallel, various forms of MTJ-based logic devices and circuits have been demonstrated, opening a possible path for spintronic ICs to expand beyond STT-MRAM.

From the system architecture perspective, STT-MRAM is particularly compelling for system-on-chip (SOC) applications. For example, STT-MRAM can serve as an embedded "nonvolatile working memory" by combining the speed and endurance of static random access memory (SRAM) or dynamic random access memory (DRAM) with the nonvolatility of FLASH or other resistive memories, such as phase change memory (PCM) and resistive memory (RRAM). In contrast to stand-alone commodity memories, each type of SOC requires a different combination of memory attributes, such as speed, energy consumption, and reliability, including cyclic endurance and retention. Embedded STT-MRAM can then be offered in a variety of macros whose designs are customized for application-specific SOC. As a challenge, however, MTJ must be integrated into logic back-end-of-line (BEOL) without disrupting standard CMOS design rules and process integration flows. Hence, driving this emerging technology beyond device and small-array engineering toward a SOC-level validation necessitates a systematic design methodology that couples MTJ physics, memory cell, array architecture, circuit design, and system architecture. In this chapter, we present such a methodology from the perspective of STT-MRAM designed for advanced CMOS, in particular targeted for low-power mobile SOC.

2.2 Basic Device Concepts

2.2.1 Magnetic Tunnel Junction (MTJ)

An MTJ, a building block for a spintronics IC, is a variable resistor that can be configured to have binary states (1 and 0) defined by two discrete resistance values. Figure 2.2 illustrates a typical MTJ film stack that essentially consists of metallic films separated by a thin tunnel barrier, most commonly MgO on the order of 1 nm in thickness. The free layer is a soft ferromagnetic metal

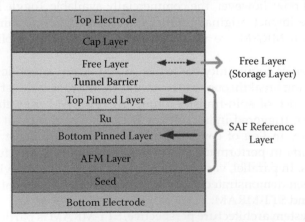

FIGURE 2.2
Illustration of a typical MTJ film stack with in-plane magnetic anisotropy.

(e.g., CoFeB, NiFe) whose magnetization can be switched by an external excitation such as spin-polarized current (i.e., STT) or current-induced magnetic field. The reference layer is commonly a synthetic structure of stacked films to provide a reference magnetization fixed in one direction (top pinned layer in Figure 2.2) relative to the free layer magnetization. The magnetization of the bottom pinned layer is fixed by an antiferromagnet (AFM) pinning layer (e.g., PtMn or IrMn) via the exchange bias effect. The top pinned layer is antiferromagnetically coupled to the bottom pinned layer via interlayer exchange coupling with a nonmagnetic spacer (e.g., Ru). This type of reference layer is called a synthetic antiferromagnetic reference layer.

The metallic films are deposited by physical vapor deposition (PVD). The MgO barrier can be grown by PVD or a combination of PVD and oxidation. MTJ device properties are tailored through a selection of desired materials and a precise control of film thickness, microstructure, and cross-sectional feature size. Key MTJ parameters, critical to optimizing performance, energy consumption, and reliability, include tunnel magnetoresistance, resistance-area product, energy barrier, and switching current, which are discussed in this chapter.

2.2.2 Tunnel Magnetoresistance

The resistance (R) of MTJ is determined by the angle (θ) between the free layer magnetization and the top pinned layer magnetization [8]:

$$R = \frac{R_\perp}{1 + \frac{TMR}{2}\cos\theta} \tag{2.1}$$

where $R_\perp$ is the resistance measured in the perpendicular magnetic configuration ($\theta = \pi/2$).

The MTJ resistance becomes minimum ($R = R_p$) for the parallel magnetization configuration ($\theta = 0$) and maximum ($R = R_{ap}$) for the antiparallel configuration ($\theta = \pi$). The tunnel magnetoresistance ratio (TMR) is then defined as

$$TMR = \frac{R_{ap} - R_p}{R_p} \times 100\% \tag{2.2}$$

Because the signal margin of sensing a dense array of MTJ (e.g., in STT-MRAM) is governed by TMR, it is critical to achieve high TMR for error-free and high-speed read operations.

TMR is a quantum mechanical phenomenon that results from spin-dependent tunneling as originally proposed by Jullière in 1975 [4]. The conductance of a metal-insulator-metal (MIM) structure is largely affected by the electron density of states near the Fermi energy level that are available for conduction electrons. When conduction electrons are emitted from one ferromagnetic metal electrode (FM2, Figure 2.1), they are spin-polarized to the magnetization direction of FM2 and tunnel through the thin tunnel barrier with their spin states conserved. Because the electron density of states in the opposite ferromagnetic metal electrode (FM1) that these tunneling electrons encounter is dependent on the magnetization direction of FM1, the conductance of the FM1-I-FM2 structure is determined by relative orientations of the magnetizations (Figure 2.1).

Since the first experimental demonstration of room temperature TMR (>10%) in 1995 [9, 10], many significant breakthroughs have been made. TMR up to ~70% was reported using an amorphous AlO_x tunnel barrier [11], which was originally adopted for conventional MRAM products. A more significant TMR breakthrough was achieved using MgO tunnel barriers. In 2001, Butler et al. [12] and Mathon and Umerski [13] theoretically predicted that extraordinarily high TMR (>1000%) could be obtained from fully epitaxial Fe(001)/MgO(001)/Fe(001) MTJs. These findings addressed the possibility of much greater electron transmission through the highly spin-polarized Δ_1 band that exists in the FM-I-FM structure, which was validated experimentally with well-crystallized MgO and high-quality interfaces [14, 15]. In 2008, experimental TMR reached ~600% at room temperature in a CoFeB/MgO/CoFeB junction [16]. To be practical, high TMR needs to be achieved in conjunction with relatively low resistance-area product (*RA*), preferably <10 Ω-cm^2, in a full-stack MTJ. However, a demonstration of such high TMR is clear evidence that desired MTJ properties can be achieved through precise materials design and process control.

From a microstructure perspective, the most critical factor in achieving high TMR is promoting strong MgO (001) texture. For CoFeB/MgO/CoFeB MTJ, it is also known that the band structure of CoFeB, modulated by the Co/Fe ratio, affects TMR considerably [17]. As-deposited CoFeB is amorphous and provides a smooth surface for MgO growth. RF-sputtered MgO promotes (001) texture in the as-deposited state, which then provides a template for CoFeB crystallization during sequential annealing [18]. A microstructure characterization of high-TMR MTJ revealed an in-plane 45° rotational epitaxial relationship between BCC-CoFe and NaCl-structured MgO. This can naturally be understood by considering the lattice constants of CoFe (0.283 ~ 0.286 nm) and MgO (0.421 nm).

An alternative way of fabricating MgO tunnel barriers is to adopt an oxidation process. For example, naturally oxidized MgO has been known to provide superior film uniformity and higher breakdown voltage for low RA MgO. Uniformity control is a difficult challenge in realizing STT-MRAM. Though oxidation was known to be a preferred method in this aspect of fabrication, it was difficult to obtain high TMR due to the poor crystallinity of oxidized MgO. The crystallization process of amorphous CoFeB was also affected by the lack of a crystallization template, resulting in a poor epitaxial relationship between the layers. Recently, this problem has successfully been resolved by inserting a thin CoFe layer between CoFeB and MgO [19]. CoFe serves as a crystallization template to induce preferred grain growth in MgO, and then to promote crystallization of CoFeB through annealing. It is remarkable that a TMR of 253% at RA = 5.9 Ω-cm^2 has been demonstrated by using such a bilayered structure. In situ annealing of the MgO barrier has been known to promote the (001) texture further, resulting in high TMR (>170%) even for MTJ films with an ultralow RA (~1 Ω-cm^2) [20]. At this time, it appears that natural oxidation is a preferred method of MgO growth for STT-MRAM.

2.2.3 Energy Barrier

At a static mode, MTJ maintains its data state without power (i.e., nonvolatile) as long as the magnetic anisotropy of its free layer is greater than the thermal excitation energy described by $k_B T$, where k_B is the Boltzmann constant and T is temperature. For MTJ with in-plane magnetic anisotropy, which is typically patterned into an elliptically shaped cell, the free layer magnetic moment can have only two energetically favorable states ($\theta = 0$ or π) along the long axis (called easy axis) of the MTJ, thereby allowing either R_p or R_{ap}. Under the assumption of a single-domain nanomagnet, the energy barrier (E_B) between these two states is often given by

$$E_B = \frac{M_s H_k V}{2}\left(1 - \frac{H_{ext}}{H_k}\right)^2 \tag{2.3}$$

where M_s is the saturation magnetization of the free layer, V is the free layer volume, H_k is the effective uniaxial anisotropy field, and H_{ext} is the external field applied along the easy axis (which vanishes in the absence of any stray field). For MTJ to be nonvolatile, E_B must be larger than the thermal excitation energy over a range of operating and storage temperatures.

The thermal disturb probability, $P(t)$, of a single MTJ is commonly described by the Néel-Brown relaxation time formula, $P(t) = 1 - exp(-t/\tau)$, with the relaxation time constant $\tau = \tau_0 exp(E_B/k_B T)$. τ_0 is ordinarily assumed to be 1 ns. For a single MTJ to retain its state for 10 years, therefore, E_B must be 40 $k_B T$ (~1 eV) or larger. Due to a statistical nature, however, the E_B requirement increases with the population of MTJ. In practice, determining a precise E_B criterion is not trivial and often requires a sophisticated statistical reliability analysis, which is eventually to be determined by application requirements and acceptable bit error rates.

To serve deeply scaled STT-MRAM, it is necessary to scale down MTJ—hence the free layer volume. Referring to Equation (2.3), this results in smaller E_B. To compensate for this loss of E_B, the M_s-H_k product needs to be increased. For an in-plane MTJ, larger M_s leads to higher switching current, consequently increasing power consumption. It is therefore more desirable to enhance H_k.

2.2.4 STT Switching

2.2.4.1 Background

A conventional way of programming MTJ is to apply an external magnetic field to switch the free layer magnetization. In fact, Toggle MRAM has utilized a combination of magnetic fields generated by currents flowing through two orthogonally aligned metal lines adjacent to the MTJ, as illustrated in Figure 2.3. A drawback of this method is the requirement of large current to induce sufficient magnetic field. It is also recognized that this method does not provide good scalability because decreasing the MTJ size entails larger switching fields, and hence even more current. At this time, scaling down this type of field-switching MRAM beyond the 90 nm node is questionable.

A breakthrough in the physics of MTJ switching was accomplished in 1996 by the theoretical formulation that the free layer magnetization could be modulated by the direct transfer of spin angular momentum from spin-polarized electrons [21, 22]. This phenomenon, called spin-transfer-torque (STT) magnetization reversal, provided a new means to control the free layer magnetization by directly applying electric current through a patterned MTJ without an external magnetic field. The magnitude of STT scales with the current density (J). This is particularly beneficial for device scalability because the critical switching current I_c should scale proportionally to the size of the MTJ.

The first experimental evidence of STT switching was discovered in 2000 in a Co/Cu/Co metallic spin valve [23]. STT switching in a discrete MTJ device was demonstrated in 2004 [24, 25]. Since then, considerable efforts have been made by academia and industry to apply this new discovery to the building

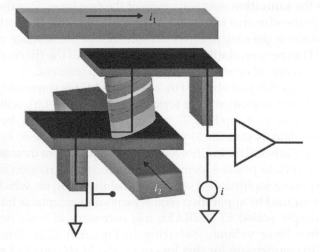

FIGURE 2.3
Schematic representation of Toggle MRAM. Each bit has an MTJ connected to an access transistor for read. The MTJ is elliptical in shape with its long axis oriented along the diagonal of the two orthogonally oriented write lines (carrying currents i_1 and i_2 to generate magnetic fields).

of a new device. Most recognized has been STT-MRAM as a next-generation memory technology. A remarkable demonstration of STT-MRAM at an array level was first reported by Hosomi et al. in 2005, including a TMR of 160% and switching speed as fast as 1 ns [26].

2.2.4.2 Switching Current: Phenomenological Description

In the case of 3d ferromagnetic metals used in typical MTJ devices, STT switching can be understood as a result of the s-d exchange interaction between spin-polarized conduction s-electrons and d-electrons responsible for magnetism. Figure 2.4 illustrates the STT switching process. When electrons flow from the fixed layer to the free layer (Figure 2.4(a)), conduction s-electrons emitted from the fixed layer are spin-polarized to the magnetization orientation of the fixed layer, tunnel through the MgO barrier without losing their polarization (majority spin), and exert torques on d-electrons in the free layer via the s-d exchange interaction. In this process, the transverse spin angular momentum of s-electrons is transferred to d-electrons as predicted by the total angular momentum conservation law. When a sufficiently large electric current is applied, its STT reverses the free layer magnetization, resulting in magnetization reversal from R_{ap} to R_p. The threshold current is called the critical switching current (I_c). For the $R_p \rightarrow R_{ap}$ switching, conduction electrons are injected into the free layer first and polarized to the magnetization orientation of the free layer (Figure 2.4(b)). In a simplified picture, minority spin electrons, polarized to the opposite direction of the free layer magnetization, are reflected at the barrier interfaces and then exert torques on the free layer magnetization.

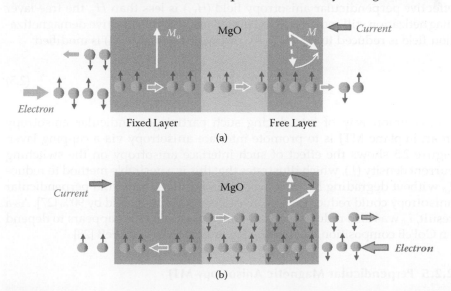

FIGURE 2.4
Simplified illustration of STT switching in MTJ: (a) $R_{ap} \rightarrow R_p$ switching and (b) $R_p \rightarrow R_{ap}$ switching.

For an in-plane MTJ, I_c at 0 K, called the intrinsic critical switching current (I_{c0}), is given by

$$I_{c0} = \frac{2e}{\hbar} \frac{\alpha}{\eta} M_s V \left(H_{k\parallel} + \frac{H_d}{2} \right) \tag{2.4}$$

where α is the damping constant, V is the free layer volume, η is the spin-torque efficiency, $H_{k\parallel}$ is the uniaxial anisotropy field in the film plane, and H_d is the effective perpendicular demagnetization field that corresponds to the field required to saturate the free layer moment perpendicular to the film plane. For typical CoFeB-based free layers, α is ~0.01 and M_s is ~1000 emu/cc. $H_{k\parallel}$ is less than 0.5 kOe. The H_d term originates from thin-film shape anisotropy. In the absence of interfacial anisotropy, H_d is given by $4\pi M_s$ ~ 12 kOe. The H_d term represents an additional energy term that needs to be overcome during STT switching because the shape anisotropy induces an oscillatory motion of magnetization confined in the direction perpendicular to the film plane, resulting in an elliptical precession. Hence, H_d only increases I_{c0} without contributing to E_B.

2.2.4.3 Interface Anisotropy

The intrinsic critical switching current (I_{c0}) is an essential figure of merit in building STT-MRAM. A primary challenge is to reduce I_{c0} without degrading E_B. Based on Equation (2.4), an effective way of doing this is to introduce perpendicular anisotropy to cancel a substantial portion of H_d. When the effective perpendicular anisotropy field ($H_{k\perp}$) is less than H_d, the free layer magnetization still remains in the film plane. But, the effective demagnetization field is reduced to $H_d - H_{k\perp}$. Accordingly, Equation (2.4) is modified:

$$I_{c0} = \frac{2e}{\hbar} \frac{\alpha}{\eta} M_s V \left(H_{k\parallel} + \frac{H_d - H_{k\perp}}{2} \right) \tag{2.5}$$

A common way of introducing such partial perpendicular anisotropy in an in-plane MTJ is to promote interface anisotropy via a capping layer. Figure 2.5 shows the effect of such interface anisotropy on the switching current density (J_c), which illustrates that this is a desirable method to reduce I_{c0} without degrading E_B. Prior work showed that this partial perpendicular anisotropy could reduce the effective demagnetization field by 90% [27]. As a result, I_{c0} was also reduced substantially. In addition, $H_{k\perp}$ appears to depend on CoFeB composition, facilitating lower I_{c0} in Fe-rich CoFeB [28].

2.2.5 Perpendicular Magnetic Anisotropy MTJ

The success of STT-MRAM is largely dependent on whether MTJ can be scaled to deep nanometer-scale nodes (20 nm and below) in conjunction

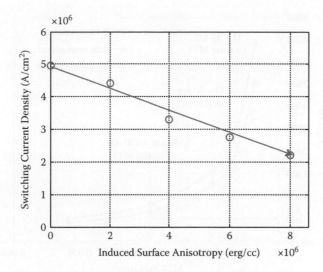

FIGURE 2.5
Critical switching current calculated as a function of induced interface anisotropy. The MTJ size is 44×124 nm with a thickness of 1.8 nm. The saturation magnetization of the free layer is 1,375 emu/cc. The current pulse width is 500 ns.

with low switching energy and high stability (adequate E_B for nonvolatility). STT-MRAM promises good scalability in terms of switching energy. However, it is challenging to achieve sufficiently large E_B (> 60 $k_B T$) as MTJ is aggressively scaled. In a macro-spin approximation, to maintain E_B, the product of uniaxial anisotropy and free layer thickness, $K_u t = (M_s t)H_k/2$, should be increased proportionally to compensate for the reduced MTJ area. For example, Figure 2.6 examines the H_k requirement for varying the MTJ area to obtain E_B of 60 $k_B T$. $M_s t$, which is a directly measurable quantity, is assumed to be 0.16, 0.2, or 0.24 memu/cm². For an in-plane MTJ, it is desirable to increase H_k because I_c is directly proportional to $M_s t$. H_k of an in-plane MTJ with a typical CoFeB-based free layer is dominated by shape anisotropy. A common way to achieve this is to increase the aspect ratio (AR) of an elliptical MTJ. However, when AR becomes too large (>3), E_B may even decrease due to a nucleation-driven magnetization reversal. Accordingly, H_k expected from shape anisotropy is on the order of 0.5 kOe, which may not be sufficient to meet the H_k requirement for MTJ areas smaller than 0.002 μm². This is equivalent to the size of an in-plane MTJ with 30 nm in short axis and AR ~ 2.8 or a circular MTJ with ~50 nm in diameter. For the MTJ area of 0.00125 μm², corresponding to a circular MTJ with its diameter of 40 nm, the H_k requirement is ~2 kOe when $M_s t$ is 0.2 memu/cm². Furthermore, elliptical MTJ shapes are more difficult to control than circular patterns. A device variability caused by this becomes particularly problematic as the MTJ size is scaled down. Therefore, from both materials and manufacturing perspectives, it does not

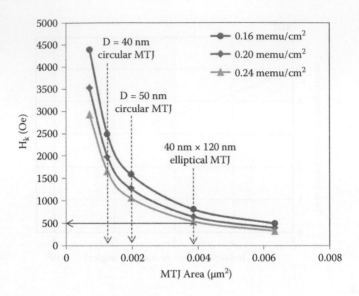

FIGURE 2.6
Effective uniaxial anisotropy (H_k) requirements as a function of MTJ area to meet E_B of 60 $k_B T$ based on a macro-spin approximation.

appear to be compelling to pursue an in-plane MTJ to produce a dense array of MTJ (e.g., >1 Gbit STT-MRAM) at deeply scaled nodes.

These challenges can be overcome by using a perpendicular MTJ (pMTJ) in which all the magnetizations in the film stack are perpendicular to the film plane. The energy barrier of a pMTJ is determined by crystalline or interface magnetic anisotropy. Hence, a pMTJ can be patterned into a circular shape. Various materials with perpendicular magnetic anisotropy (PMA) have been investigated, which include L1$_0$-ordered FePt or FePd alloys [29, 30], Co-based superlattices such as Co/Pd and Co/Pt laminates [31–34], rare earth/transition metal alloys [35, 36], etc. To build useful MTJ devices, however, these materials must be engineered for an optimal combination of materials properties like M_s and H_k and device properties like TMR and J_c. Recently, remarkable results have been reported in pMTJ materials, device engineering, and even chip-level demonstration.

A recent report has addressed that the anisotropy resulting from the CoFeB-MgO interface can induce large $H_{k\perp}$ [36]. This has been attributed to hybridization of Fe 3d and O 2p orbitals. When CoFeB is sufficiently thin (typically ~1.5 nm or thinner), such interface anisotropy can overcome the demagnetization field, that is, $H_d < H_{k\perp}$. The film can then become magnetized fully perpendicular to the plane. Perpendicular MTJ devices utilizing thin CoFeB layers have successfully been demonstrated [37–39]. The uniaxial anisotropy of MTJs with this type of interface anisotropy is provided by the effective perpendicular anisotropy. Hence, I_{c0} and E_B are described by

$$I_{c0} = \frac{e}{\hbar} \frac{\alpha}{\eta} M_s V H_{k\perp}^{eff} \qquad (2.6)$$

$$E_B = \frac{M_s V H_{k\perp}^{eff}}{2} \qquad (2.7)$$

where $H_{k\perp}^{eff}$ is the effective perpendicular anisotropy field.

In contrast to an in-plane MTJ described by Equations (2.3) to (2.5), note that I_{c0} is directly proportional to E_B. Thus, a straightforward way of reducing I_c without affecting E_B is to reduce the damping constant (α). As illustrated in Figure 2.7, the dynamics of STT switching is highly sensitive to the magnitude of α. For an α of 0.02, the magnetization starts to oscillate, and then the angle with respect to the easy axis rapidly increases, leading the gyration quickly to a complete magnetization reversal. For relatively large α of 0.05 and 0.2, the magnetization reversals do not occur and only steady magnetization gyrations at equilibrium oblique angles are maintained. In such cases, a complete magnetization reversal does not occur unless the injected current amplitude substantially increases, thereby causing large I_c. In general, α can be tuned by adjusting bulk material properties. The thickness of a free layer is also known to affect α [40]. For example, a ferromagnetic resonance measurement showed higher α for thinner CoFeB. To avoid higher α for a CoFeB-based pMTJ, therefore, it is important to secure sufficient interface PMA, which allows relatively thicker CoFeB. Furthermore, α of a thin ferromagnetic layer adjacent to a noble metal can become significantly larger due

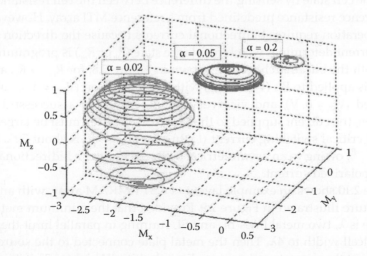

FIGURE 2.7
Simulated magnetization trajectories of a pMTJ free layer with three different damping constants.

to the spin pumping effect [41, 42]. For example, Pt, which is a strong spin sinker, substantially increases the effective damping constant of an adjacent ferromagnetic layer. Hence, it is also important to optimize the capping layer to reduce α, and hence I_c.

2.3 Device

2.3.1 Bitcell and Array

STT-MRAM is a hybrid IC built on a combination of CMOS and MTJ. Its building block is called a bitcell, which represents 1 bit. As shown in Figure 2.8, a common form of bitcell is architected in 1T-1MTJ, for which the MTJ is connected in series to an n-type metal oxide semiconductor field effect transistor (MOSFET), also called an NMOS transistor. This transistor (T) is called an access transistor as it controls read and write access to the connected MTJ as a digital switch.

Figure 2.9 is a schematic representation of a typical STT-MRAM array that consists of 1T-1MTJ bitcells. To read the information stored in a cell, the word line (WL) of the selected cell is turned on and a small read current is applied to either the selected bit line (BL) or the source line (SL) with the other end of the cell grounded. A sense amplifier, implemented by CMOS circuits, determines the cell state by sensing the difference between the cell resistance and the reference resistance predefined from a reference MTJ array. However, the write operation requires bidirectional currents because the direction of the write current determines which resistance state (R_p or R_{ap}) is programmed to MTJ. With the bitcell architecture shown in Figure 2.8, for $R_{ap} \rightarrow R_p$, a write voltage is applied to BL ($V_{BL} = V_{DD}$) with WL turned on ($V_{WL} = V_{DD}$) and SL grounded ($V_{SL} = 0$ V), and vice versa for $R_p \rightarrow R_{ap}$. For a successful write operation, the current supplied to the MTJ in each bitcell must be larger than the MTJ critical switching current (I_c). In general, $I_C^{P \rightarrow AP}$ is about 50% larger than $I_C^{AP \rightarrow P}$, owing to an asymmetrical STT effect under a bidirectional flow of spin-polarized current.

Figure 2.10 shows an example layout of a STT-MRAM array, with an array architecture illustrated in Figure 2.9. Provided that the minimum metal feature size is λ, two metal lines BL and SL running in parallel limit the minimum bitcell width to 4λ. Then the metal plate connected to the source and the drain of the access transistor may limit the bitcell height to 3λ if it is larger than 1.5 times the gate pitch along the bitcell height direction. Accordingly, the bitcell size can be as small as $12\lambda^2$.

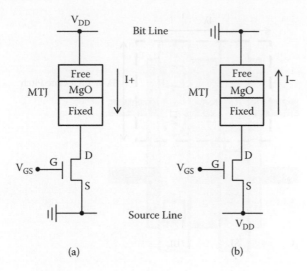

FIGURE 2.8
Schematic illustration of a conventional 1T-1MTJ bitcell: (a) Current ($_I$+) flows from bit line (BL) to source line (SL) to switch from R_{ap} to R_p and (b) current ($_I$–) flows from SL to BL to switch from R_p to R_{ap}.

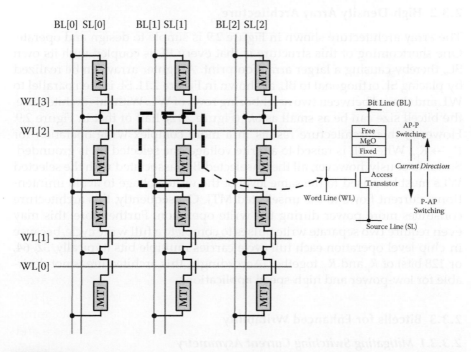

FIGURE 2.9
STT-MRAM array (4×3) with SL parallel to BL. This architecture realizes a simple bidirectional operation at low power.

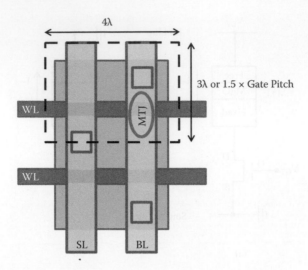

FIGURE 2.10
See color insert. A representative 1T-1MTJ bitcell for the array architecture in Figure 2.9. λ is assumed to be the minimum metal width. The bitcell size is limited by the minimum metal pitch and/or the gate pitch.

2.3.2 High-Density Array Architecture

The array architecture shown in Figure 2.9 is simple to design and operate. One shortcoming of this structure is that every BL is coupled with its own SL, thereby causing a larger array footprint. A tighter array can be realized by placing SL orthogonal to BL, as shown in Figure 2.11. SL is then parallel to WL and shared between two neighboring rows of WL. With this architecture, the bitcell size can be as small as $6\lambda^2$ (Figure 2.12), half of that in Figure 2.9. However, this architecture results in a more complex write operation for $R_p \rightarrow R_{ap}$. When SL is raised to a write voltage, the selected BL is grounded. Simultaneously, however, all the unselected BLs associated with the selected WLs must be raised to the same level of the write voltage to avoid unintentional current flows to the unselected MTJ. Consequently, this architecture consumes more power during the write operation. Furthermore, this may even require two separate write pulses to complete a full write cycle, because in chip-level operation each full cycle carries multiple bits (typically, 32, 64, or 128 bits) of R_p and R_{ap} together. Accordingly, this architecture is not desirable for low-power and high-speed applications.

2.3.3 Bitcells for Enhanced Writability

2.3.3.1 Mitigating Switching Current Asymmetry

MTJ switching is a current-induced phenomenon, and the switching operation requires a bidirectional control of current. For 1T-1MTJ, the currents

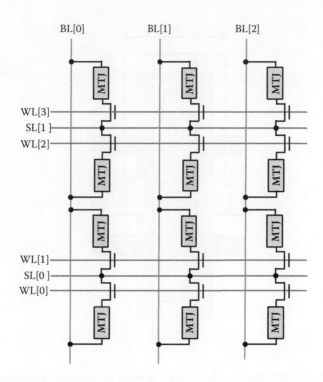

FIGURE 2.11
STT-MRAM array (4 × 3) with SL orthogonal to BL. This architecture realizes a high-density array.

supplied to MTJ are not symmetrical with respect to the polarity of current, owing to the phenomenon known as the source degeneration effect. This occurs when a resistive load is placed at the source side of a transistor. As a consequence, despite the same operating voltage (V_{DD}) applied to BL or SL, the transistor output currents are asymmetrical. This causes a significant disadvantage that reduces the write margin of 1T-1MTJ and often forces an increase in transistor size or operation voltage.

Furthermore, the STT effect on a typical MTJ is also asymmetrical, which is described by I_c asymmetry (β), defined as $\left| I_c^{P-AP} / I_c^{AP-P} \right|$. Typical MTJ devices exhibit β of 1.5 or larger, presumably due to smaller STT efficiency for $R_p \rightarrow R_{ap}$ (electrons flowing from the free layer to the reference layer). Coupling these two effects in a conventional 1T-1MTJ bitcell, it is much more difficult to switch the cell from R_p to R_{ap}, which often creates difficult limitations in cell size and operation voltage.

There are well-known approaches that can mitigate these problems: (1) a top pinned MTJ film stack and (2) a modified 1T-1MTJ with a reversely connected MTJ [43]. The first approach is to reverse the film stack sequence, commonly called a top pinned structure. The free layer is positioned toward the transistor and the reference layer is formed above the MgO barrier

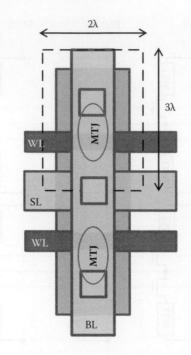

FIGURE 2.12
A representative 1T-1MTJ bitcell for the array architecture in Figure 2.11. λ is assumed to be the minimum metal width. The bitcell size is limited by the minimum metal pitch.

(Figure 2.13(a)). In this case, the switching polarity is also reversed, so that the more restrictive $R_p \rightarrow R_{ap}$ switching is not subjected to the source degeneration effect. When β is larger than the asymmetry in the transistor output currents, this approach effectively improves the write margin by aligning the asymmetric nature of the transistor and the MTJ cell. While conceptually straightforward and simple, however, this requires additional material engineering efforts because the magnetic properties of the reversed MTJ film stack, particularly the free layer, are considerably different from those of the conventional scheme (bottom pinned). The second approach is to reverse the interconnect routing between the access transistor and the MTJ (Figure 2.13(b)). Figure 2.14 shows an example that I_c can be fully covered by the supply current and the voltage provided in the reversed bitcell, whereas $R_p \rightarrow R_{ap}$ cannot be realized in the conventional bitcell. This architecture does not necessitate any change in MTJ itself, but may need more space to allow interconnect routing flexibility, resulting in a larger bitcell. When the bitcell size requirement is critical for high-density applications, this approach is not favorable. However, for embedded memory applications where a much greater bitcell size is allowable due to logic design rule constraints and performance requirements, this option may realize an increased write operation margin without causing a trade-off in bitcell size.

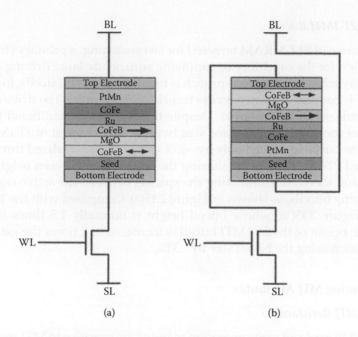

FIGURE 2.13
Alternative STT-MRAM bitcell architectures to improve write margin by reducing the coupling effect of switching current asymmetry and source degeneration effect: (a) top pinned MTJ and (b) reversely connected MTJ [1].

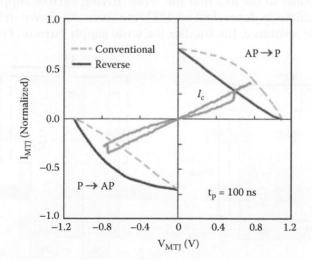

FIGURE 2.14
Load-line analysis for the MTJ switching loop that illustrates an improvement in switching margin with a reversely connected MTJ. For switching to occur, I_{MTJ} must be larger than I_c (critical switching current). For the conventional cell in this example, the $R_p \rightarrow R_{ap}$ switching cannot be realized [43].

2.3.3.2 2T-1MTJ Bitcell

In most cases of STT-MRAM targeted for fast switching, a primary challenge is to design for the capability of supplying sufficiently large driving current for MTJ switching. A simple approach is to utilize 2T-1MTJ bitcells, for which one MTJ is coupled with two access transistors in parallel. The drive current can become significantly larger. Despite the fact that an additional transistor makes the effective transistor size twice as large as that of 1T-1MTJ, the bitcell size can be increased only by ~33% to $16\lambda^2$. This is realized through an optimized 2T-1MTJ layout by sharing the source line between neighboring bitcells, and therefore eliminating the spacing between the active regions of neighboring bitcells, as shown in Figure 2.15(a). Compared with the 1T-1MTJ bitcell (Figure 2.15(b)) whose bitcell height is normally 1.5 times the gate pitch, the height of the 2T-1MTJ bitcell is increased to 2 times the gate pitch, thereby increasing the bitcell size by ~33%.

2.3.4 Tuning MTJ Attributes

2.3.4.1 MTJ Resistance

STT-MRAM read and write operation margins are sensitive to MTJ resistance. For read operations, in general, larger MTJ resistance can yield a larger sensing margin because the voltage sensing signal is proportional to MTJ resistance. For common write operations, however, smaller MTJ resistance is preferred. This is attributed to the fact that the write driving current supplied by the bitcell is significantly dependent on MTJ resistance. As shown in Figure 2.16, the larger the resistance, the smaller the write supply current. Furthermore,

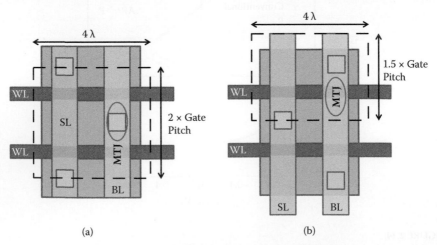

(a)   (b)

FIGURE 2.15
Comparison between (a) 2T-1MTJ and (b) 1T-1MTJ. λ is assumed to be the minimum metal width. The height of each bitcell is assumed gate pitch limited.

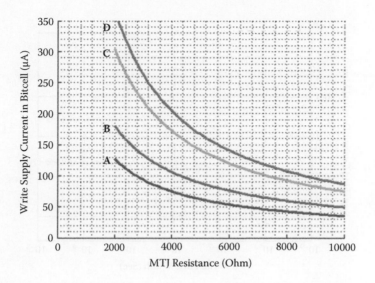

FIGURE 2.16
The write supply current as a function of MTJ resistance for conventional and reversely connected 2T-1MTJ bitcells. Curves A and D are $R_{ap} \to R_p$ and $R_p \to R_{ap}$ for the reversely connected cell, respectively. Curves B and C are for $R_p \to R_{ap}$ and $R_{ap} \to R_p$ for the conventional cell, respectively.

the source degeneration effect described in Section 2.3.3.1 severely deteriorates the bitcell write driving capability. In Figure 2.16, curves A and D represent the supply currents for $R_{ap} \to R_p$ and $R_p \to R_{ap}$, respectively, in the case of a reversely connected bitcell (Figure 2.13(b)). In comparison, curves B and C stand for the supply current for $R_p \to R_{ap}$ and $R_{ap} \to R_p$, respectively, in the case of a conventional bitcell (Figure 2.8). For a reversely connected bitcell, $R_{ap} \to R_p$ has a limited write driving capability, whereas for a conventional bitcell, $R_p \to R_{ap}$ is the limited case. Provided that transistor attributes (e.g., channel length and width, threshold voltage) are given, whether to choose a conventional or a reversely connected bitcell largely depends on β (I_c asymmetry). In general, when β is larger than the asymmetry in the transistor output currents, the write margin is effectively improved by compromising the asymmetric natures of the transistor and the MTJ cell.

2.3.4.2 MTJ Write Speed

In STT-MRAM, I_c has a strong dependence on write pulse width, as illustrated in Figure 2.17. Fast MTJ switching, often referred to as precessional switching (~10 ns or below), requires substantially larger I_c than relatively slow switching. This leads to difficult challenges in designing high-performance bitcells. Unless the MTJ size is substantially small, it is difficult to realize a practical bitcell without enlarging the bitcell size (often painfully). This is a

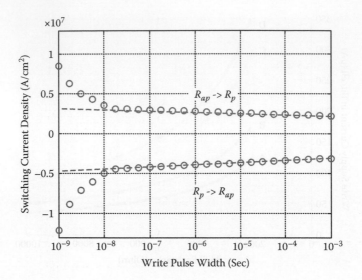

FIGURE 2.17
Switching current density as a function of current pulse width for the two switching paths ($R_{ap} \rightarrow R_p$ and $R_p \rightarrow R_{ap}$).

primary reason why continuing innovations in MTJ materials engineering are still necessary to reduce J_c (critical switching current density).

Accordingly, it is desired to tune MTJ and bitcell attributes tailored for varying write speed requirements depending on different STT-MRAM product applications. For example, for embedded level 2 or 3 CPU cache memory, the MTJ switching speed needs to be on the order of a few nanoseconds, although this could often be relaxed significantly through various design optimization techniques. In contrast, for traditional embedded nonvolatile memory applications, a switching speed on the order of a microsecond is still compelling (a few orders of magnitude faster). An advantage of STT-MRAM is such that MTJ can be tuned for custom bitcells that can meet widely varying ranges of product applications.

2.3.4.3 Effective TMR

TMR is an important MTJ parameter that affects the MTJ read operation margin. In a current sensing circuitry, the ratio of a signal current with respect to a reference current can be expressed as

$$\frac{I_{signal}}{I_{ref}} = \frac{TMR^{eff}}{TMR^{eff} + 2} \tag{2.8}$$

Here TMR^{eff} is the effective TMR between R_p and R_{ap}, which is smaller than TMR of MTJ itself owing to the inclusion of additional components along the read circuit path. Since TMR^{eff} is substantially smaller than 2, the sensing margin is nearly proportional to

$$\frac{TMR^{eff}}{2}.$$

Hence, read speed and margin can be enhanced by increasing TMR^{eff}, that is, by increasing MTJ TMR and also optimizing the read path.

2.3.4.4 Energy Barrier: Data Retention

As described in Section 2.2.3, E_B is a critical parameter for MTJ thermal stability, and thereby static data retention of stored bits in a STT-MRAM array. Figure 2.18 shows a distribution of calculated retention time for a 1 Mbit STT-MRAM array as a function of E_B in order to achieve 99% chip yield. Note that E_B here represents 25°C, and that it tends to decrease with temperature. For consumer applications, it is generally considered that $E_B \approx 60\,k_BT$ is adequate enough to satisfy a typical 10-year data retention requirement. In comparison, for high-temperature applications such as automobile microcontrollers, which often require such long data retention at 125°C and fault-free operation even at 150°C, E_B must be $\sim 80\,k_BT$ or greater.

The challenge of achieving large E_B while maintaining low I_c is acute. This challenge can be mitigated by the addition of an error correction code (ECC). For example, a common single-bit ECC can reduce the E_B requirement from

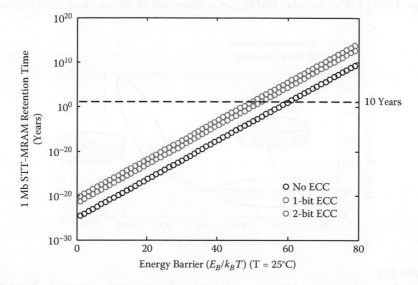

FIGURE 2.18
Energy barrier (E_B) requirements for data retention to achieve 99% chip yield for 1 Mb STT-MRAM.

60 k_BT to 52 k_BT for the case examined in Figure 2.18. A more conservative double-bit ECC can further decrease the E_B requirement to 49 k_BT, however, with an increased chip-area overhead and performance penalty.

As described in Equation (2.3), in-plane MTJ relies on the shape anisotropy to maintain two different magnetization states. Any abnormality in shape or even edge roughness, which may be introduced during MTJ fabrication, can negatively impact E_B. Figure 2.19 illustrates E_B, calculated by micromagnetic modeling [44], by deliberately creating edge roughness as much as 4 nm. Also considered here is the effect of induced perpendicular anisotropy (Sections 2.2.4.2 and 2.2.5). Such edge roughness can reduce E_B by nearly 15 k_BT when the anisotropy largely remains in-plane. This is particularly a concern for tail bits that suffer from much larger parametric variations, causing problems including serious losses in data retention, as can be expected from Figure 2.18. It is surprising, however, that when the induced anisotropy turns the MTJ into fully perpendicular, the edge roughness effect disappears.

Referring to Equations (2.3) and (2.7), E_B scales down as the MTJ size is reduced. Assuming ordinary conditions of H_k and M_s in Equation (2.3), the simulation results in Figure 2.20 suggest that in-plane MTJ can be scaled down to ~45 nm (short axis) with an inclusion of edge roughness. Below ~45 nm, in-plane MTJ should face a difficult challenge to achieve adequate E_B unless further innovations in materials are achieved to allow larger H_k and M_s while I_c is still tolerable.

E_B of perpendicular MTJ (pMTJ) shall not be governed by MTJ shape (Figure 2.19). Figure 2.21 compares E_B for a circularly shaped and an elliptically shaped pMTJ. The circular pMTJ has a diameter of 28 nm, and the elliptical

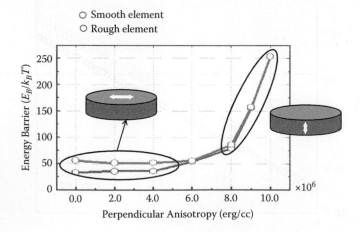

FIGURE 2.19

Energy barrier (E_B) calculated as a function of induced perpendicular anisotropy. When the perpendicular anisotropy exceeds 6×10^6 erg/cc, the magnetization in MTJ turns into perpendicular from in-plane. Two types of MTJ (smooth edge versus rough edge) are compared. The MTJ size is kept as 49×106 nm and the free layer thickness is 1.8 nm.

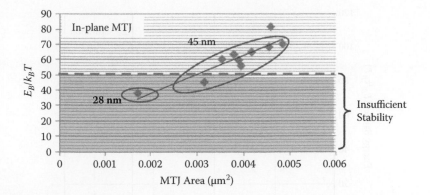

FIGURE 2.20
Energy barrier (E_B) of in-plane MTJ calculated as a function of MTJ area. Two sets of MTJ are compared with the short axes of 28 and 45 nm. The aspect ratios are varied from 2.2 to 3.0. The free layer is 1.8 nm thick CoFeB.

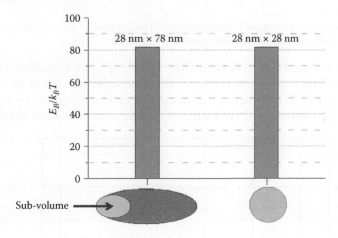

FIGURE 2.21
Energy barrier (E_B) simulated for pMTJ with an elliptical shape of 28×78 nm and a circular shape of 28 nm in diameter.

pMTJ is almost three times larger, with the size of 28×78 nm. However, the E_B values are nearly identical for these two cases. This is presumably attributed to the possibility that the switching mode of perpendicular MTJ dynamically limits the thermal stability of MTJ larger than a sub-volume, as illustrated in Figure 2.21. This subject will be discussed in Section 2.4.2.2.

A significant implication of the sub-volume-dominated magnetization reversal is that when MTJ size exceeds a critical value, E_B may no longer increase with the MTJ size. This is illustrated in Figure 2.22, where a 9 Å thick CoFeB free layer is assumed to have a perpendicular anisotropy of 1.0×10^7 erg/cc. In this simulation, E_B remains essentially unchanged for MTJ diameters above ~34 nm. Below this, E_B monotonically decreases as the

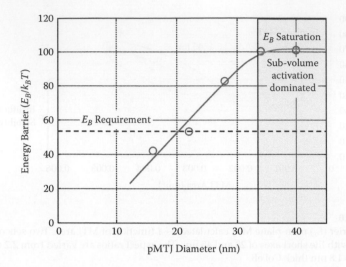

FIGURE 2.22
Energy barrier (E_B) as a function of pMTJ diameter. The free layer is assumed to be 0.9 nm thick with an induced perpendicular anisotropy of 1.0×10^7 erg/cc.

MTJ size is reduced. This suggests that achieving E_B for sufficient data retention is a challenge for such a thin CoFeB free layer at ~20 nm or below. As shown in Figure 2.17, however, this challenge can be mitigated by tuning the perpendicular anisotropy. For example, the simulation results in Figure 2.23

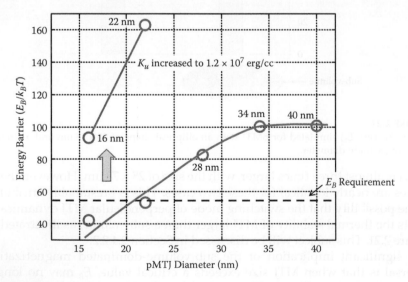

FIGURE 2.23
Energy barrier (E_B) as a function of pMTJ diameter, indicating that the increase in perpendicular anisotropy from 1.0×10^7 to 1.2×10^7 erg/cc can increase the energy barrier substantially to ensure adequate E_B for deeply scaled pMTJ.

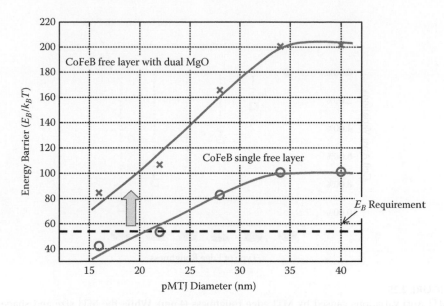

FIGURE 2.24
Energy barrier (E_B) enhancement through adopting CoFeB-based dual-MgO interface pMTJ.

indicate that E_B can increase significantly by slightly increasing the perpendicular anisotropy from 1.0×10^7 to 1.2×10^7 erg/cc.

There are a variety of known material systems with high perpendicular anisotropy. For example, FePt can achieve anisotropy on the order of $\sim 7 \times 10^7$ erg/cc. The availability of such high-anisotropy materials should enable the scaling of pMTJ well below 10 nm, provided that lithography is not a limiting factor. Another well-known solution to enhance E_B is to engineer the free layer with dual-MgO interfaces. Figure 2.24 shows that E_B can be doubled provided that the free layer thickness is increased twice while the perpendicular anisotropy is maintained. It should be noted that these approaches do not necessarily cause J_c to rise, for example, by tuning materials properties such as damping and saturation magnetization.

2.4 Variability

2.4.1 Sources of Variations

STT-MRAM is poised for superior scalability compared with conventional charge-based memories. However, STT-MRAM also is not immune from common causes of process variations owing to aggressive scaling in MTJ feature size coupled with difficulties in manufacturing control.

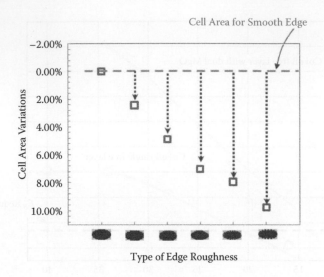

FIGURE 2.25
MTJ area variations caused by MTJ edge roughness (4 nm). While the MTJ size and shape appear equivalent, the area can vary as much as 10%.

The first type of variability is known as systematic variability. Modern semiconductor fabrication processes are operated in a subwavelength optical lithography regime, leading to large critical dimension (CD) variations. Figure 2.25 illustrates potential MTJ area variations in the presence of MTJ edge roughness. Despite the fact that all these MTJs appear to be equivalent in physical size and shape, small edge roughness (~4 nm for Figure 2.25), which may be introduced during MTJ fabrication, can lead to MTJ area variations as large as 10%. This type of CD variation can then cause significant variations in MTJ resistance and switching current.

In order to reduce MTJ CD variations, advanced reticle enhancement techniques, such as optical proximity correction, phase shift masks, and for a half pitch smaller than ~30 nm, even double patterning, may be applied. Furthermore, several critical process modules, such as MTJ etching and chemical mechanical polishing, are sensitive to device pattern density. Optimizing pattern densities and also making them uniform can reduce MTJ line edge roughness and also film interface roughness. These types of systematic variability necessitate a design-for-manufacturing methodology tailored for MTJ.

The second type of variability is random stochastic variability. In CMOS, it is well recognized that random dopant fluctuations may result in serious transistor mismatches. This problem can also affect STT-MRAM by causing large variability in access NMOS, and consequently in both writability (drive current variation) and readability (sensing margin). In addition, there are various extrinsic sources of statistical variability introduced during the manufacturing process that are difficult to predict. Thus, it is important to understand the impact of random variability and to design for it.

2.4.2 Switching Current Variation

The dynamics of STT switching are complex. Largely a nondeterministic switching process can cause MTJ to suffer from a stochastic random variation in I_c [45].

2.4.2.1 STT Switching in In-Plane MTJ

STT switching is precessional in nature. Figure 2.26 shows a transient switching process of an element of $160 \times 80 \text{ nm}^2$ with the averaged magnetization components plotted as a function of time (the x- and y-axes are along the long and short axes of the elliptical element, respectively). The switching does not occur by following a deterministic path. Rather, the averaged magnetization components indicate that the dynamic switching is a random oscillatory process. As the switching progresses, the magnetization precessional angles with respect to the long axis of the MTJ increase, until the magnetization becomes completely reversed. Figure 2.27 shows a snapshot of a transient magnetization configuration during switching, with the color contrast representing the perpendicular component of the magnetization and the arrows standing for the in-plane components of magnetization. A complex high-order spin wave mode is generated along both axes. Not only is the spatial

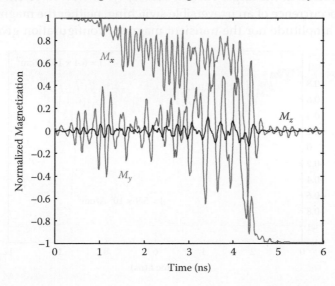

FIGURE 2.26
Volume-averaged magnetization components simulated as a function of time for an in-plane MTJ of $160 \times 80 \text{ nm}^2$ [45].

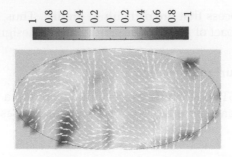

FIGURE 2.27

See color insert. Transient magnetization configurations during STT switching for an in-plane MTJ of 160 × 80 nm. The color represents the M_z component.

wavelength of the spin wave a small fraction of the element size, but also the configuration appears to be irregular and random.

The nonlinear high-order spin wave mode may cause the magnetization switching to be probabilistic if the current amplitude is not sufficiently large. Figure 2.28 shows an example case, which plots the volume-averaged magnetization of the storage layer during the injection of a current pulse at two different current densities. Both in-plane magnetization components (M_x and M_y) appear to be irregular, which causes the final magnetization state to become uncertain. At $J = 5.9 \times 10^6$ A/cm^2, the magnetization has switched at the end of the current pulse, while no irreversible magnetization switching has occurred at the larger current density (J) of 6.4×10^6 A/cm^2. Prior to the actual occurrence of an irreversible switching, neither the magnetization precession amplitude nor the transient magnetic configuration gives a sign

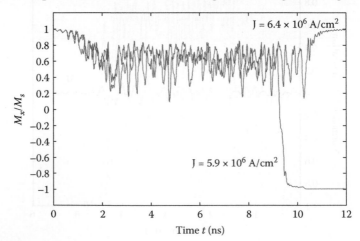

FIGURE 2.28

Volume-averaged transient in-plane magnetization components simulated for the 10 ns pulse at two different current densities. The MTJ size is 220 × 110 nm.

to predict when the irreversible switching will actually occur. If current is abruptly turned off before reaching the critical switching time, the magnetization should return to its original state. For a fixed current pulse width, this uncertainty in switching time is manifested as current amplitude fluctuation.

The transient configurations of the complicated high-order spin waves excited during STT switching are also sensitive to the initial magnetization configuration. This is attributed to the fact that for an in-plane MTJ element, the demagnetization field varies significantly from the element edge to the center, causing the initial precession frequency and amplitude to vary across the element. Combined with the effect of ferromagnetic exchange coupling within the element, the spin waves with high-order modes are generated. This excitation of high-order spin waves during STT switching is therefore inherent for in-plane MTJ.

A more serious consequence of the nonuniform transient spin waves is formation of a vortex in the middle of the MTJ. Such a vortex may remain metastable as an intermediate state. Figure 2.29 shows a vortex configuration formed at the end of an intended STT switching cycle. This type of vortex formation is a probabilistic event, and it can occur over a wide range of current amplitudes. The probability of vortex formation becomes greater for an element with a larger width or a larger thickness due to the increase in magnetostatic energy. Once a stable vortex is formed to create an intermediate state, I_c, to complete the switching, it will become substantially higher than a normal value.

The probabilistic STT switching caused by high-order spin waves can cause both I_c and the switching time to fluctuate over a relatively wide range. Consequently, at finite current pulses, the resulting magnetization state will become uncertain over a broad range of current amplitudes, resulting in I_c variations. The transient configuration of a high-order spin wave has been determined by the combination of two competing effects: the magnetostatic energy, which drives the nonuniform switching and effectively reduces shape anisotropy, and the ferromagnetic exchange coupling trying to keep magnetization uniform. At small MTJ sizes, the relatively strong ferromagnetic exchange energy promotes a coherent magnetization rotation of the entire storage layer. Hence, the distribution of

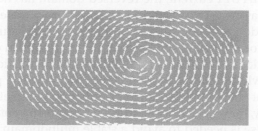

FIGURE 2.29
See color insert. Formation of a magnetization vortex at the end of the STT switching process.

switching currents is much narrower. To estimate the MTJ size of coherent magnetization switching, the characteristic exchange length in the free layer is defined as $L_{ex} = \sqrt{A / K_{shape}}$, where $K_{shape} \approx 10^5$ erg/cc estimated from the shape anisotropy field. To eliminate the formation of high-order spin waves during switching, the MTJ size needs to be less than the exchange length $L_{ex} \approx 30$ nm. Although I_c of in-plane MTJ could be suppressed by reducing the element size, the required thermal stability (E_B) may ultimately limit the size reduction.

2.4.2.2 STT Switching in Perpendicular MTJ

In contrast to in-plane MTJ, STT switching in pMTJ is much more deterministic, as can be seen in Figure 2.30. The simulated MTJ has a circular shape with a diameter of 80 nm. The volume-averaged magnetization of the free layer during the injection of a current pulse is illustrated in Figure 2.30(a). The corresponding transient magnetization configurations are shown in Figure 2.30(b). The color contrast indicates the perpendicular magnetization component M_z, and the arrows plot the in-plane components of magnetization vectors. The switching starts as the magnetization curls around the element center, similar to the classic curling mode. As the switching progresses, the middle of the magnetization curling structure moves toward the element edge, because the exchange energy for the curling center at the edge is significantly lower than that at the center. The magnetization angle with respect to the initial perpendicular direction increases, and the magnetization precession becomes more and more nonuniform spatially. Consequently, the magnetization of a small area at the edge of the element is first reversed. The reversed domain then expands through the entire element as the switching process completes.

For a given current pulse duration, an insufficient STT current could leave the element in a multidomain state before the switching is complete. Figure 2.31 shows the volume-averaged perpendicular magnetization component as a function of time for a current pulse of 5 ns at a series of different amplitudes for a pMTJ (80 nm in diameter). At the end of the current pulse duration of 200 µA STT current, the reversed domain has not yet completed its expansion, leaving a two-domain configuration in the element. In practice, the formed domain wall, such as the one shown in Figure 2.30(b), could be pinned by a local pinning site, causing the two-domain configuration to be stable. If the domain wall is not pinned, the magnetization configuration should drift slowly to a single perpendicular domain state prevailed by the larger domain at the end of the current pulse. To complete the switching deterministically, however, either a sufficiently high current or a sufficiently long current pulse is needed to avoid a stable multidomain state in perpendicular MTJ. As shown in Figure 2.31(b), which is the case for a 10 ns pulse width of 200 µA switching current (versus 5 ns in Figure 2.31(a)), this longer

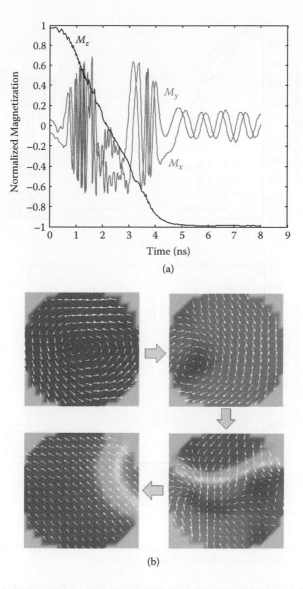

(a)

(b)

FIGURE 2.30
See color insert. (a) Simulated volume-averaged magnetization components as a function of time for a pMTJ (80 nm in diameter). (b) The corresponding transient magnetization configurations during STT switching. The color represents the M_z component [45].

current pulse can yield a complete magnetization reversal at the end of the current pulse of an intermediate amplitude.

In contrast to in-plane MTJ, the onset stage of the STT switching process in pMTJ is described by a simple and well-defined curling mode. The absence of a high-order spin wave excitation is mainly due to the cylindrical symmetry.

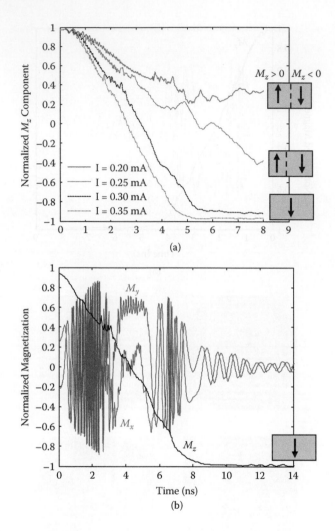

FIGURE 2.31
Simulated volume-averaged magnetization components as a function of time for a pMTJ
(80 nm in diameter) at pulse widths of (a) 5 ns and (b) 10 ns.

A curling-mode-like precession is naturally suited to the circular geometry of
the element so that there is no need to excite a high-order spin wave that is at a
higher energy state. In addition, the magnetization precession does not introduce an additional demagnetization field component as it does in the in-plane
case. Accordingly, increasing the precession amplitude only yields a reduction
in the perpendicular demagnetization field, consequently further reducing
the spatial nonuniformity. Therefore, for pMTJ, the dynamic switching process is deterministic and repeatable. This implies that the switching current
distribution pertaining to pMTJ, if significant, would result primarily from
cell-to-cell material property variations or pinning site characteristics.

2.4.3 Temperature and Voltage Variations

Temperature and operating voltage are also key factors that influence the variation in both CMOS and MTJ. For CMOS, as temperature increases, the threshold voltage (V_t) increases, resulting in the decrease of current driving capability for MTJ switching. Furthermore, MTJ characteristics are also a function of temperature, affecting both read and write operations. As shown in Figure 2.32, R_{ap} has a significant dependence on temperature, whereas R_p remains essentially constant. This causes TMR to decrease significantly with temperature. When temperature increases from −30 to 125°C, TMR can be reduced by more than 30%, which reduces a read margin at 125°C. In contrast, I_c is reduced with temperature, leading to more favorable write operations at higher temperatures. These are primarily attributed to two factors. First, STT switching is assisted by increased thermal activation. Second, lower R_{ap} allows the bitcell circuitry to supply more drive current for $R_{ap} \rightarrow R_p$.

In addition, Figure 2.32 shows that R_{ap} decreases with MTJ bias voltage. This works favorably for write ($R_{ap} \rightarrow R_p$), but not for read, owing to reduced TMR at a practical read voltage (0.1 ~ 0.2 V).

2.4.4 Design for Variability

Process, voltage, and temperature variations can cause a wide distribution of MTJ resistance, TMR, and I_c. Such variations can lead to unreliable STT-MRAM design. Higher degrees of variability might even cause a device to be discarded. In order to overcome this challenge, a new design methodology needs to be built on a robust model.

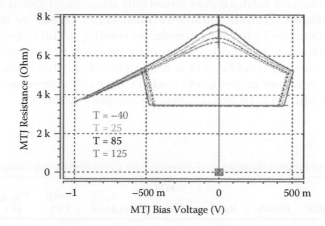

FIGURE 2.32
MTJ resistance profiles as a function of MTJ bias voltage at varying temperatures from −40° to 125°C. R_{ap} decreases as the temperature rises.

2.4.4.1 Corner Model: Worst-Case Design

To cope with systematic and random variations, a corner model, as shown in Table 2.1, can be constructed by covering boundary conditions of related device parameters. These corners are provided to represent the ranges of process, voltage, and temperature variations that a circuit designer must account for. The temperature range is usually from −30 to 125°C. The voltage range is from 10% over to 10% under target V_{DD}. The process variation range is less straightforward to define and requires a validation based on a sufficient amount of measurement data.

It is not surprising to project significantly different results in bitcell functionality when a corner model like that shown in Table 2.1 is applied. Figure 2.33 shows an example to compare the STT-MRAM switching characteristics at the fast, slow, and typical corners. Not only does I_c of MTJ vary, but the switching voltage also varies significantly. The switching voltage of the slow corner is nearly twice as large as that of the fast corner, particularly for $R_{ap} \rightarrow R_p$. This is attributed to the fact that the slow corner condition suffers from an even more serious source degeneration effect by superposing the boundary values of several parameters. Accordingly, the write circuit needs to be designed properly with an adequate margin to contain the slow corner. At a bitcell level, this affects the type of access transistor, as shown in Figure 2.33. By comparing the slower corners of two cases, it is clear that adopting low-threshold-voltage (LVt) NMOS can secure a much improved write window than standard-threshold-voltage (SVt) NMOS. Lowering the write voltage is particularly important for embedded STT-MRAM because it is desirable to match the write voltage and CMOS logic V_{DD} (i.e., no need for a charge pump).

While a corner model is common and simple to adopt, it brings shortcomings and limitations. First, a corner model only accounts for global variations, but not for local variations. Parametric mismatches caused by local variation are a dominant factor, for example, for sensing circuit design. Second, a corner model may be considered conservative, but is not realistic, because it superposes corner conditions of several parameters. This is illustrated in Figure 2.34. A corner model likely leads to an overly conservative design, causing STT-MRAM to be less competitive and pricy to build.

TABLE 2.1

An Example of Process, Voltage, and Temperature Conditions for a Corner Model

	NMOS	PMOS	MTJ Read	MTJ Write	MTJ Breakdown	Temp (°C)	$V_{|BL-SL|}$ (V)	V_{WL} (V)
Typical	T (0)	T (0)	T (0)	T (0)	T (0)	25	1.1	1.54
Fast	F (+3σ)	F (+3σ)	F (+3σ)	F (+3σ)	F (+3σ)	125	1.21	1.65
Slow	S (−3σ)	S (−3σ)	S (−3σ)	S (−3σ)	S (−3σ)	−30	0.99	1.43

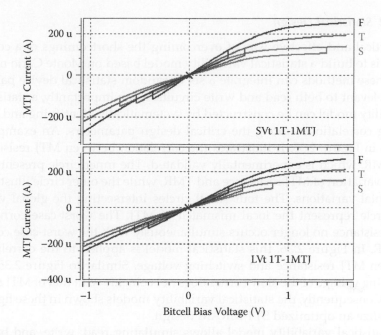

FIGURE 2.33

MTJ switching curves at the corner conditions described in Table 2.1. To secure switching at the worst-case write condition, the SVt case needs SL overdrive. The LVt case has an adequate margin without SL or BL overdrive. F, T, and S stand for the fast, the typical, and the slow corner, respectively.

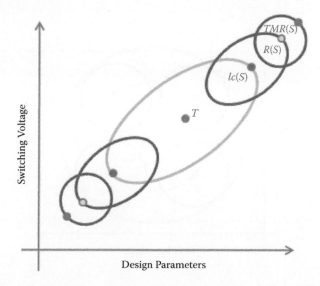

FIGURE 2.34

Illustration of a corner model design methodology. The worst-case switching voltage is projected upon superposing worst-case parameters of MTJ switching current, resistance, and TMR.

2.4.4.2 Statistical Design

A practical and effective way of overcoming the shortcomings of a corner model is to build a statistical variability model based on Monte Carlo methods. These methods can integrate a set of random statistical device parameters relevant to both read and write circuits. More importantly, a statistical variability model can be constructed by systematically identifying and integrating correlations among the critical design parameters. An example is shown in Figure 2.35. It illustrates the correlation between MTJ resistance and TMR, which is experimentally validated. The inner circle presents the global variation of MTJ resistance and TMR, while the outer circle illustrates their total variations. The four small circles intersecting the global variation circle represent the local mismatch of MTJ. The worst-case corner of MTJ resistance no longer occurs simultaneously with the worst-case corner of TMR. In Figure 2.36, this statistical model is applied to the correlation between MTJ resistance and switching voltage. Similar to Figure 2.35, the switching voltage of the slow corner does not coincide with that of MTJ resistance. Consequently, the statistical variability models shown in these figures can realize an optimized STT-MRAM design.

A statistical variability model allows simulating read, write, and breakdown voltages at the chip level. An example is shown in Figure 2.37. Because MTJ read and write operations are performed by applying a bias voltage across MTJ, it is important to secure a sufficient separation (Δ_1) between read and write voltages. Furthermore, the MTJ voltage must not exceed a threshold

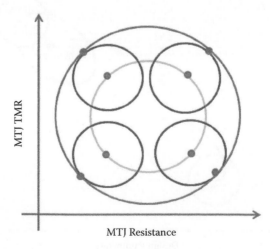

FIGURE 2.35
Illustration of the correlations between MTJ resistance and TMR over an STT-MRAM array. Understanding correlations among physical parameters like these is essential to build a statistical variability model (versus a corner model).

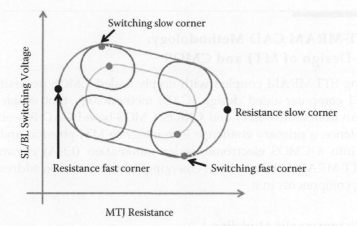

FIGURE 2.36
Illustration of a statistical design methodology plotting the relationship between bitcell switching voltage and MTJ resistance. The worst-case switching voltage does not coincide with the worst corner of MTJ resistance.

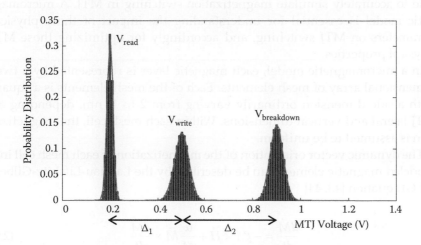

FIGURE 2.37
Design windows defined by MTJ read, write, and breakdown voltage distributions.

value above which the tunnel barrier is subjected to time-dependent dielectric breakdown, requiring also an adequate separation (Δ_2) between MTJ switching and breakdown voltages. To achieve a functional and reliable chip, STT-MRAM design must ensure, at least, 12σ separations among the three distributions shown in Figure 2.37.

2.5 STT-MRAM CAD Methodology: Co-Design of MTJ and CMOS

Designing STT-MRAM coupled with deeply scaled CMOS necessitates an advanced computer-aided design (CAD) methodology that enables a co-design environment of MTJ and CMOS. CMOS-based CAD is well established. Hence, a primary challenge is to integrate MTJ physical and circuit models into a CMOS electronic design automation (EDA) environment. While STT-MRAM CAD is still an emerging field, this section addresses key enabling components in it.

2.5.1 Micromagnetic Modeling

A micromagnetic model simulates and analyzes the underlying device physics of MTJ. Many physical parameters such as MTJ size, aspect ratio, roughness, shape, and materials properties directly affect STT-MRAM switching performance. It has been widely recognized that a micromagnetic model is able to accurately simulate magnetization switching in MTJ. A micromagnetic model is essential for understanding the impact of those physical parameters on MTJ switching, and accordingly for optimizing those MTJ physical properties.

In a micromagnetic model, each magnetic layer is represented by a two-dimensional array of mesh elements. Each of the mesh elements is a square with a side dimension ordinarily varying from 2 to 10 nm, depending on MTJ lateral and vertical dimensions. Within each mesh cell, the magnetization is assumed to be uniform.

The dynamic vector orientation of the magnetization of each mesh cell in a modeled magnetic element can be described by the Landau-Lifshitz-Gilbert (LLG) equation [43, 44]:

$$\frac{d\vec{M}}{dt} = -\gamma\vec{M} \times \vec{H} + \frac{\alpha}{M}\vec{M} \times \frac{d\vec{M}}{dt} \tag{2.9}$$

where $\vec{M}$ is the magnetization, γ is the electron gyromagnetic ratio, $\vec{H}$ is the effective magnetic field, and α is the phenomenological damping constant. The micromagnetic model developed in this way can successfully describe the two parts of the magnetization motion during its dynamic switching process. One part is the gyromagnetic motion without energy dissipation, and the other part is the magnetization damping process toward the direction of the effective magnetic field. For STT switching, the effect of STT is added into the LLG equation as an additional torque term, leading to the STT-modified Gilbert equation:

$$\frac{d\vec{M}}{dt} = -\gamma\vec{M} \times \vec{H} + \gamma\frac{P_0 J\hbar}{eM\delta} \vec{M} \times \widehat{M_0 \times \hat{M}} + \frac{\alpha}{M} \vec{M} \times \frac{d\vec{M}}{dt} \qquad (2.10)$$

where the second term on the right-hand side of the equation represents STT, with P_0 and M_0 denoting the polarization factor and the polarization direction of the spin current, respectively, δ is the magnetic layer thickness, J is the current density, $\vec{H}$ is the effective magnetic field, and $\vec{M}$ is the local saturation magnetization. Note that the effective magnetic field is defined as the negative of the energy gradient with respect to magnetization [48]:

$$\vec{H} = -\frac{\partial E}{\partial \vec{M}} \qquad (2.11)$$

where the energy is the sum of magnetostatic energy (often referred to as demagnetization energy), exchange energy, anisotropy energy, and magnetic potential energy due to the external magnetic field (also known as Zeeman energy).

Furthermore, it is also important to include a thermal effect in the model, as thermal energy can agitate magnetization fluctuations in MTJ, which can assist STT switching. The thermally activated local magnetization fluctuation can be modeled with a random magnetic field that is added into the effective magnetic field $\vec{H}$. The thermal field is assumed to be random both in direction and in magnitude. It follows a Gaussian distribution with its variance given by the fluctuation-dissipation theorem [49, 50]. Each random thermal field is applied over a finite time.

A micromagnetic model can simulate STT-MRAM switching behaviors at various conditions. The results are valuable in closely linking MTJ physical properties to bitcell and circuit operating conditions (voltage, current, temperature, switching speed, etc.).

2.5.2 MTJ Compact Model: HSPICE Model

STT-MRAM CAD necessitates inclusion of a MTJ compact model that can seamlessly be integrated into a common CMOS circuit simulation environment such as HSPICE. An MTJ compact model simulates the electrical behavior of MTJ. MTJ is a variable resistor that can be configured to have binary states (0 and 1) defined by two discrete resistance values (R_p and R_{ap}, respectively). Hence, it can be modeled as a voltage-controlled variable resistor. Figure 2.38 is a schematic illustration of a basic circuit implementation that emulates MTJ electrical characteristics. This equivalent circuit is composed of two variable resistors; one of the two is controlled by a switch. When the switch is on, a low-resistance state (R_p) is set, whereas a high-resistance stage (R_{ap}) is established otherwise. The resistor parallel to the switch is controlled

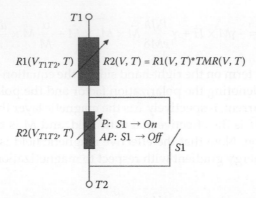

FIGURE 2.38
Schematic description of a MTJ circuit model applied to build an MTJ HSPICE compact model.

by the voltage between the two terminals of the circuit, which models R_{ap} that depends on the voltage across MTJ (e.g., Figure 2.32).

An MTJ compact model can simulate MTJ switching hysteresis loops. Figure 2.39 includes both the I-V loop (Figure 2.39(a)) and the R-V loop (Figure 2.39(b)). STT switching occurs when the current flowing through MTJ exceeds I_c, an essential model parameter that is predetermined by device measurement results. In the model, I_c is correlated to the switching voltage (V_c)

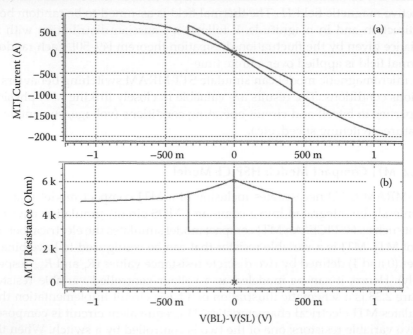

FIGURE 2.39
MTJ switching hysteresis loops obtained by an MTJ compact model: (a) I-V and (b) R-V.

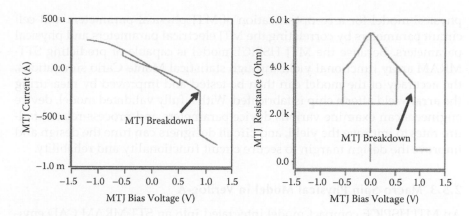

FIGURE 2.40
MTJ electrical behaviors simulated by an MTJ compact model with an inclusion of MTJ breakdown: (a) I-V and (b) R-V.

through MTJ resistance (R_p and R_{ap}). A MTJ compact model needs to include a capability simulating a MTJ breakdown behavior, as shown in Figure 2.40. When the applied voltage across MTJ exceeds a breakdown threshold value, the model outputs an abrupt drop in resistance because there is no effective tunnel barrier. Also observed is an abrupt increase in MTJ current.

A co-design methodology that integrates an MTJ compact model into a CMOS-based HSPICE environment can be described by the flowchart in Figure 2.41. Utilizing this HSPICE model, all necessary STT-MRAM device parameters are evaluated and tuned to meet a target yield of STT-MRAM array. The HSPICE model can then be combined with a micromagnetic

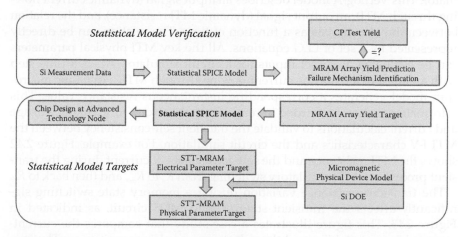

FIGURE 2.41
A flowchart illustrating an STT-MRAM co-design methodology by incorporating a statistical MTJ compact model into a CMOS-based EDA environment. The validity of the methodology can be checked by chip-level measurement data.

physical model for a co-optimization of MTJ physical parameters and cell circuit parameters by correlating the MTJ electrical parameters and physical parameters. Because the MTJ HSPICE model is capable of predicting STT-MRAM array functional yield through statistical Monte Carlo simulations, the accuracy of the model can then be tested and improved by measuring the array yield after a chip is fabricated. With a fully validated model, device engineers can examine various device parameters and process-related failure rates to improve the yield, and circuit designers can tune the design and improve the design margin to secure circuit functionality and reliability.

2.5.3 Macro-Spin Physical Model in Verilog-A

An MTJ HSPICE compact model integrated into an STT-MRAM CAD environment is essentially an empirical model obtained by fitting experimentally measured data. Owing to the limitation that device testing can measure only static electrical switching phenomena of an STT-MRAM cell, this empirical HSPICE model emulates only a static switching behavior between the two resistance states (R_p and R_{ap}), similar to a step function. In reality, however, MTJ switching is a dynamic process for which the transient-state resistance is not monotonic but oscillatory. Accordingly, the transient I-V behavior may not be ignored in circuit simulations. In order to include a capability of simulating dynamic MTJ behavioral characteristics coupled with CMOS logic circuits, an MTJ macro-spin model can be implemented by integrating a physical dynamic model into a circuit simulation environment.

In general, an MTJ physical model is implemented at a circuit level using an analog modeling language, Verilog-A, which is compatible with a SPICE simulator. This Verilog-A model describes an input signal (dynamic current flowing through MTJ), an output signal (dynamic MTJ resistance), and the relation between the two signals as a function of time. The relation can be directly represented by a set of LLG equations. All the key MTJ physical parameters can be evaluated as direct inputs for a circuit simulator. At each time step with an input of a real-time current flowing through MTJ, a dynamic macro-spin model generates a corresponding instantaneous resistance value. This information is then forwarded in real time to the circuit for dynamic voltage and current calculations to validate the transient self-consistency between the MTJ I-V characteristics and the circuit simulation. For example, Figure 2.42 shows the MTJ resistance and the self-consistency of current during the transient process of a MTJ oscillatory switching from R_p to R_{ap}, and then back to R_p.

The transient resistance variation during a memory state switching significantly affects the transient state of the CMOS circuit, as indicated in Figure 2.43. This figure illustrates a case study that compares the current-source-driven switching and the voltage-source-driven switching. The current-source-driven switching is equivalent to a static switching in MTJ, as the switching current is not affected by MTJ resistance variations. As a consequence, the current-source-driven switching optimistically underestimates

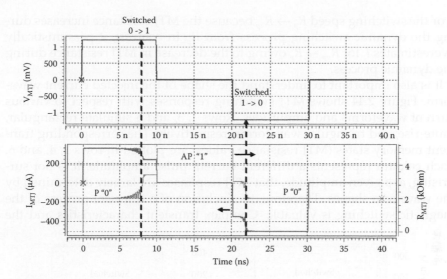

FIGURE 2.42

An example of the transient process of an MTJ oscillatory switching from R_p to R_{ap}, then back to R_p, modeled utilizing Verilog-A.

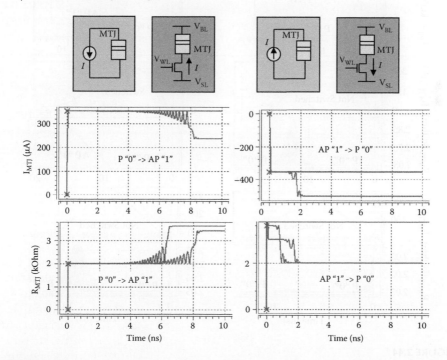

FIGURE 2.43

Comparison of dynamic switching processes between a current-source-driven MTJ (representing a conventional model) and a voltage-source-driven MTJ (representing a macro-spin model in Verilog-A). MTJ currents and resistances are plotted as a function of switching time.

I_c or the switching speed $R_p \rightarrow R_{ap}$ because the MTJ resistance increases during the dynamic switching process from R_p. In contrast, it pessimistically overestimates I_c for $R_{ap} \rightarrow R_p$ owing to the decrease in MTJ resistance during the dynamic process.

It is also important to understand the shape of the injected current waveform. Figure 2.44 shows MTJ switching responses with respect to various current waveforms and amplitudes. Rows 1, 3, and 5 simulate rectangular, finite-rise, and triangular waveforms, respectively. The corresponding transient memory states (MTJ resistance values) are plotted in rows 2, 4, and 6. Each column represents a different current pulse amplitude. It is not surprising, at a given amplitude, that MTJ responses are largely determined by the waveform shapes. Therefore, the inclusion of a transient process of the magnetic switching is valuable. Complex transient characteristics and the

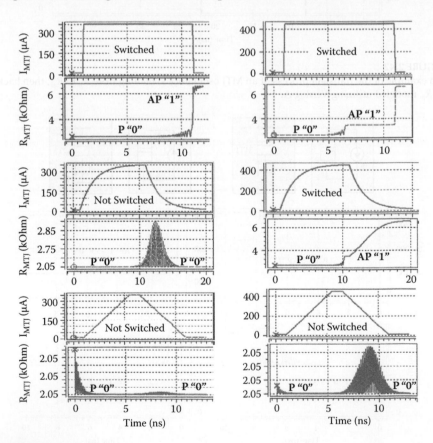

FIGURE 2.44
Current waveform effects on memory state switching. Rows 1, 3, and 5 are the injected current pulses of rectangular, finite-rise, and triangular waveforms, respectively. The corresponding transient changes of the memory states, i.e., MTJ resistance values, are plotted as a function of time in rows 2, 4, and 6. Different columns correspond to different pulse amplitudes.

real-time self-consistence can be obtained to enable a simultaneous optimization of MTJ and CMOS logic circuits.

For simplicity and compatibility with Verilog-A, a macro-spin model is built on the assumption that each memory state is represented by a uniform magnetization. In reality, however, spin states are complex (Figure 2.27), so that micromagnetic modeling is necessary, for example, to describe high-order spin wave excitations in in-plane MTJ or a formation of transient multi-domain states in pMTJ. Furthermore, this macro-spin model implemented in Verilog-A is relatively slow to execute, practically limiting its value for chip-level design applications. Including a demand for a more advanced CAD method with a built-in micromagnetic model, STT-MRAM CAD remains an emerging field to advance further with continuing development efforts.

References

1. Kang, S. H., K. Lee. 2013. Emerging materials and devices in spintronic integrated circuits for energy-smart mobile computing and connectivity. *Acta Materialia* 61:952–973.
2. Baibich, M. N., J. M. Broto, A. Fert, et al. 1988. Giant magnetoresistance of (001) Fe/(001)Cr magnetic superlattices. *Phys. Rev. Lett.* 61(21):2472–2475.
3. Binasch, G., P. Grünberg, F. Saurenbach, W. Zinn. 1989. Enhanced magneto-resistance in layered magnetic structures with antiferromagnetic interlayer exchange. *Phys. Rev. B* 39(7):4828–4830.
4. Jullière, M. 1975. Tunneling between ferromagnetic films. *Phys. Lett.* 54A:225–226.
5. Durlam, M., B. Craigo, M. DeHerrera, et al. 2007. Toggle MRAM: A highly-reliable non-volatile memory. In *International Symposium on VLSI Technology, Systems and Applications (VLSI-TSA 2007)*, pp. 1–2.
6. Savtchenko, L., B. N. Engel, N. D. Rizzo, M. F. Deherrera, J. A. Janesky. 2003. Method of writing to scalable magnetoresistance random access memory element. U.S. Patent 6,545,906.
7. Engel, B. N., J. Åkerman, B. Butcher, et al. 2005. A 4-Mbit toggle MRAM based on a novel bit and switching method. *IEEE Trans. Magn.* 41:132–136.
8. Jaffrès, H., D. Lacour, F. Nguyen Van Dau, J. Briatico, F. Petroff, A. Vaurès. 2001. Angular dependence of the tunnel magnetoresistance in transition-metal-based junctions. *Phys. Rev. B* 64:064427.
9. Miyazaki, T., N. Tezuka. 1995. Giant magnetic tunneling effect in Fe/Al$_2$O$_3$/Fe junction. *J Magn. Magn. Mater.* 139:L231–L234.
10. Moodera, J. S., L. R. Kinder, T. M. Wong, R. Meservey. 1995. Large magneto-resistance at room temperature in ferromagnetic thin film tunnel junctions. *Phys. Rev. Lett.* 74:3273–3276.
11. Wei, H. X., Q. H. Qin, M. Ma, R. Sharif, X. F. Han. 2007. 80% tunneling magnetoresistance at room temperature for thin Al–O barrier magnetic tunnel junction with CoFeB as free and reference layers. *J. Appl. Phys.* 101:09B501.

12. Butler, W. H., X. G. Zhang, T. C. Schulthess, J. M. MacLaren. 2001. Reduction of electron tunneling current due to lateral variation of the wave function. *Phys. Rev. B* 63(5):054416.

13. Mathon, J., A. Umerski. 2001. Theory of tunneling magnetoresistance of an epitaxial Fe/MgO/Fe (001) junction. *Phys. Rev. B* 63(22):220403.

14. Yuasa, S., T. Nagahama, A. Fukushima, Y. Suzuki, K. Ando. 2004. Giant room-temperature magnetoresistance in single-crystal Fe/MgO/Fe magnetic tunnel junctions. *Nat. Mater.* 3:868–871.

15. Parkin, S. S. P., C. Kaiser, A. Panchula, P. M. Rice, B. Hughes, M. Samant, S.-H. Yang. 2004. Giant tunnelling magnetoresistance at room temperature with MgO (100) tunnel barriers. *Nat. Mater.* 3:862–867.

16. Ikeda, S., J. Hayakawa, Y. Ashizawa, et al. 2008. Tunnel magnetoresistance of 604% at 300 K by suppression of Ta diffusion in CoFeB/MgO/CoFeB pseudo-spin-valves annealed at high temperature. *Appl. Phys. Lett.* 93:082508.

17. Tsunekawa, K., Y.-S. Choi, Y. Nagamine, D. D. Djayaprawira, T. Takeuchi, Y. Kitamoto. 2006. Influence of chemical composition of CoFeB on tunneling magnetoresistance and microstructure in polycrystalline CoFeB/MgO/CoFeB magnetic tunnel junctions. *Jpn. J. Appl. Phys.* 45:L1152–L1155.

18. Hayakawa, J., S. Ikeda, F. Matsukura, H. Takahashi, H. Ohno. 2005. Dependence of giant tunnel magnetoresistance of sputtered CoFeB/MgO/CoFeB magnetic tunnel junctions on MgO barrier thickness and annealing temperature. *Jpn. J. Appl. Phys.* 44:L587–L589.

19. Choi, Y.-S., H. Tsunematsu, S. Yamagata, H. Okuyama, Y. Nagamine, K. Tsunekawa. 2009. Novel stack structure of magnetic tunnel junction with MgO tunnel barrier prepared by oxidation methods: Preferred grain growth promotion seed layers and bi-layered pinned layer. *Jpn. J. Appl. Phys.* 48:120214.

20. Maehara, H., K. Nishimura, Y. Nagamine, et al. 2011. Tunnel magnetoresistance above 170% and resistance–area product of 1 Ω (μm)^2 attained by in situ annealing of ultra-thin MgO tunnel barrier. *Appl. Phys. Exp.* 4:033002.

21. Berger, L. 1996. Emission of spin waves by a magnetic multilayer traversed by a current. *Phys. Rev. B* 54:9353–9358.

22. Slonczewski, J. C. 1996. Current-driven excitation of magnetic multilayers. *J. Magn. Magn. Mater.* 159:L1–L7.

23. Katine, J. A., F. J. Albert, R. A. Buhrman, E. B. Myers, D. C. Ralph. 2000. Current-driven magnetization reversal and spin-wave excitations in Co/Cu/Co pillars. *Phys. Rev. Lett.* 84:3149–3152.

24. Fuchs, G. D., N. C. Emley, I. N. Krivorotov, et al. 2004. Spin-transfer effects in nanoscale magnetic tunnel junctions. *Appl. Phys. Lett.* 85:1205–1207.

25. Huai, Y., F. Albert, P. Nguyen, M. Pakala, T. Valet. 2004. Observation of spin-transfer switching in deep submicron-sized and low-resistance magnetic tunnel junctions. *Appl. Phys. Lett.* 84:3118–3120.

26. Hosomi, M., H. Yamagishi, T. Yamamoto, et al. 2005. A novel nonvolatile memory with spin torque transfer magnetization switching: spin-RAM. In *IEEE International Electron Devices Meeting 2005, IEDM Technical Digest*, pp. 459–462.

27. Driskill-Smith, A., D. Apalkov, V. Nikitin, et al. 2011. Latest advances and roadmap for in-plane and perpendicular STT-RAM. In *3rd IEEE International Memory Workshop (IMW)*, pp. 1–3.

28. Yakata, S., H. Kubota, Y. Suzuki, et al. 2009. Influence of perpendicular magnetic anisotropy on spin-transfer switching current in CoFeB/MgO/CoFeB magnetic tunnel junctions. *J. Appl. Phys.* 105:07D131.

29. Yoshikawa, M., E. Kitagawa, T. Nagase, et al. 2008. Tunnel magnetoresistance over 100% in MgO-based magnetic tunnel junction films with perpendicular magnetic $L1_0$-FePt electrodes. *IEEE Trans. Magn.* 44:2573–2576.

30. Khan, M. N. I., N. Inami, H. Naganuma, Y. Ohdaira, M. Oogane, Y. Ando. 2012. Promotion of $L1_0$ ordering of FePd films with amorphous CoFeB thin interlayer. *J. Appl. Phys.* 111:07C112.

31. Mizunuma, K., S. Ikeda, J. H. Park, et al. 2009. MgO barrier-perpendicular magnetic tunnel junctions with CoFe/Pd multilayers and ferromagnetic insertion layers. *Appl. Phys. Lett.* 95:232516.

32. Yakushiji, K., T. Saruya, H. Kubota, et al. 2010. Ultrathin Co/Pt and Co/Pd superlattice films for MgO-based perpendicular magnetic tunnel junctions. *Appl. Phys. Lett.* 97:232508.

33. Markou, A., I. Panagiotopoulos, T. Bakas, et al. 2011. Formation of $L1_0$ with (001) texture in magnetically annealed Co/Pt multilayers. *J. Appl. Phys.* 110:083903.

34. Rahman, M. T., A. Lyle, G. Hu, W. J. Gallagher, J.-P. Wang. 2011. High temperature annealing stability of magnetic properties in MgO-based perpendicular magnetic tunnel junction stacks with CoFeB polarizing layer. *J. Appl. Phys.* 109:07C709.

35. Ohmori, H., T. Hatori, S. Nakagawa. 2008. Perpendicular magnetic tunnel junction with tunneling magnetoresistance ratio of 64% using MgO (100) barrier layer prepared at room temperature. *J. Appl. Phys.* 103:07A911.

36. Yakushiji, K., K. Noma, T. Saruya, et al. 2010. High magnetoresistance ratio and low resistance–area product in magnetic tunnel junctions with perpendicularly magnetized electrodes. *Appl. Phys. Exp.* 3:053003.

37. Ikeda, S., K. Miura, H. Yamamoto, et al. 2010. A perpendicular-anisotropy CoFeB–MgO magnetic tunnel junction. *Nat. Mater.* 9:721–724.

38. Worledge, D. C., G. Hu, D. W. Abraham, et al. 2011. Spin torque switching of perpendicular Ta|CoFeB|MgO-based magnetic tunnel junctions. *Appl. Phys. Lett.* 98:022501.

39. Guenole, J., Y.-J. Wang, T. Moriyama, Y.-J. Lee, M. Lin, T. Zhong, R.-Y. Tong, T. Torng, P.-K. Wang. 2012. High spin torque efficiency of magnetic tunnel junctions with MgO/CoFeB/MgO free layer. *Appl. Phys. Exp.* 5:093008.

40. Bilzer, C., T. Devolder, J.-V. Kim, et al. 2006. Study of the dynamic magnetic properties of soft CoFeB films. *J. Appl. Phys.* 100:053903.

41. Tserkovnyak, Y., A. Brataas, G. E. W. Bauer. 2002. Enhanced Gilbert damping in thin ferromagnetic films. *Phys. Rev. Lett.* 88:117601.

42. Mosendz, O., J. E. Pearson, F. Y. Fradin, S. D. Bader, A. Hoffmann. 2010. Suppression of spin-pumping by a MgO tunnel-barrier. *Appl. Phys. Lett.* 96:022502.

43. Lin, C.J., S.H. Kang, Y.J. Wang, K. Lee, X. Zhu, W.C. Chen et al. 2009. 45NM low power CMOS logic compatible embedded STT MRAM utilizing a reverse-connection IT/1MTS cell. IEEE International Electron Devices Meeting, Baltimore.

44. Zhu, X., S.H. Kang, 2012. On the thermal stability of STT-MRAM designs. HB-08. IEEE International Magnetics Conference, Vancouver.

45. Zhu, X., S.H. Kang. 2009. Inherent spin transfer torque driven switching current fluctuations in magnetic element with in-plane magnetization and comparison to perpendicular design. *J. Appl. Phys.* 106:113906.

46. Gilbert, T. L. 1955. A Lagrangian formulation of the gyromagnetic equation of the magnetization fields. *Phys. Rev.* 100:1243.
47. Gilbert, T. L. 2004. A phenomenological theory of damping in ferromagnetic materials. *IEEE Trans. Magn.* 40(6):3443–3449.
48. Zhu, J.-G. 1989. Interactive in magnetic thin films. Ph.D. thesis, University of California at San Diego, San Diego, CA.
49. Brown, W. F. 1963. Thermal fluctuations of a single-domain particle. *Phys. Rev.* 130:1677–1686.
50. Zhu, J.-G. 2002. Thermal magnetic noise and spectra in spin valve heads. *J. Appl. Phys.* 91:7273–7275.

3

The Prospect of STT-RAM Scaling

Yaojun Zhang, Wujie Wen, Hai Li, and Yiran Chen

Department of Electric and Computer Engineering, Swanson School of Engineering, University of Pittsburgh, Pittsburgh, Pennsylvania

CONTENTS

3.1 Introduction

Spin-transfer torque random access memory (STT-RAM) is an emerging nonvolatile memory technology aimed at embedded memory and on-chip cache applications. In an STT-RAM cell, data are stored as the two or more resistance states of a magnetic tunneling junction (MTJ) device, as shown in Figure 3.1(a). The resistance states of the MTJ are determined by the magnetization of the magnetic layers, which can be changed by passing through an electrical current with different polarizations. A popular one-transistor-one-MTJ (1T1J) STT-RAM cell structure is shown in Figure 3.1(d), where the NMOS transistor size must be large enough to supply sufficient switching current to the MTJ. The unique storage mechanism offers STT-RAM many attractive characteristics, such as fast operation time, small memory cell size, radiation hardness, good complementary metal oxide semiconductor (CMOS) process compatibility and scalability, etc. [21].

As technology scales, tremendous efforts have been made at reducing the MTJ switching current and improving the energy and performance of STT-RAM writes. Many new MTJ devices, for example, the MTJ with perpendicular magnetization [12] (see Figure 3.1(b)), dual tunneling barrier [18] (see

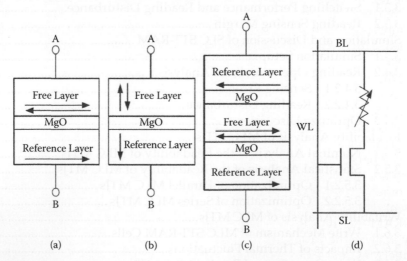

FIGURE 3.1
(a) Conventional MTJ. (b) Perpendicular MTJ. (c) Dual-barrier layer MTJ. (d) 1T1J STT-RAM cell structure.

Figure 3.1(c)), speculating layer, etc., have emerged. These new device structures achieved one order of magnitude reduction on the switching current compared to conventional in-plane MTJs. The recent experimental results show that the STT-RAM write access time can be also improved as fast as 3 ns [20]. However, the MTJ device variability incurred in the manufacturing results in the variation of the sensing margin, leading to a relatively long sensing time (e.g., longer than 2 ns) when the sensing margin is small. A long sensing time may induce read disturbance issues when the magnetization state of the MTJ is flipped during the read. As the MTJ switching current reduces, the read current or the sensing time must be scaled accordingly to maintain a low read disturbance rate. In other words, the conflict between the read performance and the read stability emerges as a critical issue following the scaling of MTJ devices.

Because the MTJ switching current in an STT-RAM cell is supplied by the MOS device, the magnitude of the switching current determines not only the write energy of STT-RAM, but also its memory integration density. Besides reducing the write current, multilevel cell (MLC) technology is also explored in STT-RAM design to store more than one data bit in a single STT-RAM cell: following the improvement on the distinction between the lowest and the highest resistance states of the MTJ, it becomes possible to further divide the MTJ resistance into multiple levels to represent the combinations of multiple data bits [7, 13]. However, compared to single-level cell (SLC) STT-RAM designs, the impacts of such design variabilities in MLC STT-RAM designs are even more prominent due to the scaled data storage margin. In general, the variability sources in STT-RAM designs include (1) the device parametric deviations of the MOS transistor and MTJ, such as the variations of the geometry sizes [5], the threshold voltage [26], and the magnetic materials [19]; and (2) the thermal fluctuations in the MTJ switching [20]. Although the impacts of these variations on the SLC STT-RAM designs have been well studied in many previous works [16], it is still unclear if MLC STT-RAM is a viable technology when all the design variabilities are taken into account. Also, the robustness of the different types of MLC MTJ designs requires further investigation because obviously their resilience to the variations varies due to the device structures' differences.

In this chapter, we first quantitatively analyze the robustness of the STT-RAM read operations tracking down the current scaling path of MTJ devices, that is, from 45 nm to 22 nm. Both persistent errors and nonpersistent errors, which are induced by process variations and thermal-induced MTJ switching randomness, respectively, are included in our simulations. We also present the importance of finding the optimal read current to enhance the STT-RAM read reliability at different technology nodes. We then systematically analyze the variation sources of two MLC STT-RAM designs, namely, parallel MLC and series MLC, and investigate the impacts of these variations on the memory performance and reliability. Our work will try to answer the questions that STT-RAM researchers have had for a long time: Can MLC STT-RAM designs be realized by using the existing technologies? And if not, how far are we from it? Luckily, our analysis shows that at least the series MLC STT-RAM may potentially be implemented by using the stacked MTJ structure [13] and achieve an acceptable reliability for commercial applications. Based on our analysis, we also discuss the design optimization methods for the minimization of the operation error rate of MLC STT-RAM.

3.2 Preliminary

3.2.1 Basic MTJ Structure

The data storage device in an STT-RAM cell is the magnetic tunneling junction (MTJ), where a tunneling oxide layer is sandwiched between two ferromagnetic layers. The MTJ resistance is determined by the relative magnetization directions of the two ferromagnetic layers: when their magnetization directions are parallel (antiparallel), the MTJ is in its low- (high-) resistance state, as shown in Figure 3.2(a). The magnetization direction of the reference layer is fixed, while that of the free layer can be flipped by passing a spin-polarized current [21]. For example, when the write current is applied from the free layer side to the reference layer side, the MTJ is programmed to the low-resistance state. If the write current is applied from the other direction, the

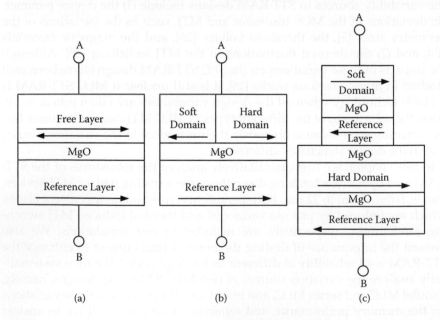

FIGURE 3.2
(a) Conventional MTJ. (b) Parallel MLC MTJ. (c) Series MLC MTJ.

MTJ is programmed to the high-resistance state. A parameter called tunneling magnetoresistance ratio (TMR) is introduced to measure the distinction between the two resistance states of the MTJ as $(R_H - R_L)/R_L$. Here R_H and R_L denote the high- and low-resistance states of the SLC MTJ, respectively.

There are two general approaches to improve the STT-RAM write performance: increasing the MTJ driving current and decreasing the MTJ threshold switching current (for a fixed switching time). As an efficient way to raise the MTJ driving current, increasing the NMOS transistor size inevitably results in STT-RAM cell area overhead and degrades the integration density. In the magnetism society, the research mainly focuses on the reduction of the MTJ threshold switching current. The intrinsic threshold current density J_{C0} of conventional in-plane MTJ devices (see Figure 3.1(a)), which is the minimum current required to flip the MTJ resistance in the absence of any external magnetic field at 0 K, is given by [11]

$$J_{C0} = \left(\frac{2e}{\hbar}\right)\left(\frac{\alpha}{\eta}\right)(t_F M_s)(H_k \pm H_{ext} + 2\pi Ms). \qquad (3.1)$$

Here e is the electron charge, α is the damping constant, Ms is the saturation magnetization, t_F is the thickness of the free layer, $\hbar$ is the reduced Planck's constant, H_k is the effective anisotropy field including magnetocrystalline anisotropy and shape anisotropy, and H_{ext} is the external field. The critical current I_{C0} should be present as $I_{C0} = J_{C0} \cdot A_{MTJ}$, where A_{MTJ} is the cross-section area of the MTJ elliptic cylinder. I_{C0} is determined by the material property of MTJ (i.e., M_s, spin efficiency η, damping ratio α, etc.), the MTJ geometry property (i.e., A_{MTJ}, t_F), and the working environment (such as working temperature). Many new MTJ device structures have emerged to reduce I_{C0} by optimizing these parameters. Some examples are:

Perpendicular MTJ: Perpendicular MTJ devices (see Figure 3.1(b)) include a free layer with a perpendicular anisotropy of $H_k > 4\pi M_s$ [11]. Thus, the H_k term in Equation (3.1) is transformed to the effective perpendicular anisotropy as $H_k^{\perp} = H_{k\perp} - 4\pi M_s$. In this way, I_{C0} can potentially be further scaled because the $2\pi M_s$ factor from the out-of-plane demagnetizing field is canceled by the perpendicular spin-transfer torque. In the meantime, other material property parameters, especially spin-transfer torque efficiency μ and damping α, maintain comparable to those in in-plane MTJ [2]. Hence, compared to the current STT-RAM implemented with in-plane MTJs, the perpendicular MTJ-based STT-RAM (P-STT) can achieve lower write energy or better write performance, while maintaining similar thermal stability.

Dual tunneling barrier MTJ: As the name indicates, a dual tunneling barrier MTJ device (see Figure 3.1(c)) is designed by adding an additional fixed magnetic layer on top of the free layer as an additional source of spin-polarized electrons [9]. The two antiparalleled fixed layers will enhance the effective spin torque magnitude on the free layer under the same switching current. In the other words, it will cost less current to flip the MTJ magnetization state, compared to the conventional MTJ structure.

3.2.2 MLC MTJ

The multilevel cell (MLC) capability can be implemented by realizing four or more resistance levels in MTJ designs. At least two proposals of MLC MTJ structures have emerged [13, 17] so far, including parallel MLC MTJs and series MLC MTJs. Figure 3.2(b) shows a two-bit parallel MLC MTJ where the free layer of the MLC MTJ has two magnetic domain whose magnetization directions can be changed separately. The partitioning of the free layer affects the shape anisotropy of each magnetic domains and results in the different switching current requirements of each domain. Therefore, the two magnetic domains, namely, soft domain or hard domain, can be switched by either a low switching current or a high switching current, respectively. Figure 3.2(c) shows a two-bit series MLC MTJ in which two single-level cell (SLC) MTJs with different sizes are serially stacked. The required switching current and the resistance of each MTJ are mainly determined by the surface area of the MTJ.

In parallel MLC MTJs, the four resistance states—00, 01, 10, and 11—are uniquely defined by the four combinations of the magnetic directions of the two magnetic domains in the free layer. The first and second digits of the two-bit data refer to the resistance states of the hard domain and the soft domain, respectively, as shown in Figure 3.3 [6]. In series MLC MTJs, the four resistance states are uniquely defined by the combinations of the relative magnetizations of the two SLC MTJs. The minimal device size of a parallel MLC MTJ and the small SLC MTJ in a series MLC MTJ can be the

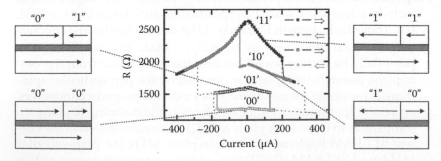

FIGURE 3.3
Four resistance states of MTJ and R-I swap curve. (From Y. Chen, Access Scheme of Multi-Level Cell Spin-Transfer Torque Random Access Memory and Its Optimization, in *53rd IEEE International Midwest Symposium on Circuits and Systems*, August 2010, pp. 1109–1112. With permission.)

same as that of the normal SLC MTJ, which is defined by the required aspect ratio and the lithography limit. We note that the parallel MLC MTJ design is only applicable to in-plane MTJ technology because it requires the different aspect ratios of the two magnetic domains to achieve different switching current densities. The series MLC MTJ design, however, is compatible with the advanced MTJ technologies, such as perpendicular MTJ, etc. [22].

3.2.3 Variability Sources in STT-RAM Designs

The performance and reliability of SLC and MLC STT-RAM cells are seriously affected by mainly two types of variabilities: (1) the process variations of MOS and MTJ devices and (2) the thermal fluctuations in MTJ switching process.

3.2.3.1 Process Variations

The major sources of MTJ device variations mainly include (1) MTJ shape variations, that is, the surface area variation; (2) MgO layer thickness variations; and (3) normally distributed localized fluctuation of magnetic anisotropy: $K = M_s \cdot H_k$. Here H_k, the effective anisotropy field, includes magnetocrystalline anisotropy and shape anisotropy. These factors lead to the deviations of MTJ resistance and the required switching current from the nominal values.

The MTJ device variations affect the reliability of the two types of MLC MTJs in different ways: in parallel MLC MTJs, the two parts of the MTJ with different magnetic domains (for simplicity, we call them "two magnetic domains" in the rest of this chapter) share the same free layer, reference layer, and MgO layer. In such a small geometry size, we can assume the MgO layer thickness and the resistance-area (RA) of these two parts are fully correlated. Other parameters, such as the MTJ surface areas, the magnetic anisotropy, and the required switching current density, can be very different for these two parts because they are determined by the magnetic domain partitioning. In series MLC MTJs, however, all these parameters of two SLC MTJs are close to each other but only spatially correlated.

We note that the MOS device variations also affect the robustness of MLC STT-RAM designs by causing the magnitude variations of the read and write currents of the MTJ. In our reliability analysis of MLC STT-RAM, the parametric variability of MOS devices is represented by the variations of the current source output.

3.2.3.2 Thermal Fluctuations

The thermal fluctuations result in the randomness of the MTJ switching time. In general, the impact of thermal fluctuations can be modeled by a normalized thermal-induced random field $\vec{h}_{fluc}$ in the stochastic Landau-Lifshitz-Gilbert (LLG) equation (Equation (3.2)) [3, 8, 10] as

$$\frac{d\vec{m}}{dt} = -\vec{m} \times (\vec{h}_{eff} + \vec{h}_{fluc})$$

$$+\alpha\vec{m} \times \left(\vec{m} \times (\vec{h}_{eff} + \vec{h}_{fluc})\right) + \frac{\vec{T}_{norm}}{M_s}. \tag{3.2}$$

Here $\vec{m}$ is the normalized magnetization vector, t is the normalized time, α is the LLG damping parameter, $\vec{h}_{eff}$ is the normalized effective magnetic field, and $\vec{T}_{norm}$ is the spin torque term with units of magnetic field. The MTJ switching time becomes a distribution under the impact of thermal fluctuations. A write failure occurs when the MTJ switching time is longer than the write pulse width. The impact of thermal fluctuations is an accumulative effect and determined by the length of the MTJ switching time. The reduction of switching current not only prolongs the MTJ switching time, but also increases the ratio between the standard deviation and the mean value of the switching time [8], indicating a larger impact of thermal fluctuations. Hence, in MLC STT-RAM designs, the impacts of thermal fluctuations could be stronger than those in the SLC STT-RAM designs when the MTJ switching current density is lower than that of the SLC MTJ (e.g., during the soft domain flipping in parallel MLC MTJs).

3.3 SLC STT-RAM Reading Performance with Technology Scaling

3.3.1 Switching Performance and Reading Disturbance

The invention of the emerging MTJ device structures leads to a fast switching performance down to a few nanoseconds [14]. However, the reduction of the MTJ threshold switching current also makes the magnetization state of MTJ easier to flip during the read operations.

The MTJ switching probability (P_{sw}) is a function of the MTJ critical switching current I_{C0}, the inverse of the attempt frequency τ_0, the applied current I_C, and the pulse width τ_p as

$$P_{sw} = 1 - \exp\left\{-\tau_p / \tau_0 \exp\left[-E/k_B T (1 - I_C/I_{C0})\right]\right\} \tag{3.3}$$

We simulated the switching probability of an elliptical MTJ with the size of 45×90 nm, as shown in Figure 3.4. To minimize the read disturbance during the read operations, the read current I_{read} must be sufficiently lower than the normal write current (which is about 100 μA for a write pulse width of 10 ns). We note that considering the randomness incurred by the device parametric variations and thermal fluctuations, the actual I_{read} may need to be set much lower than the one calculated by Equation (3.3), as

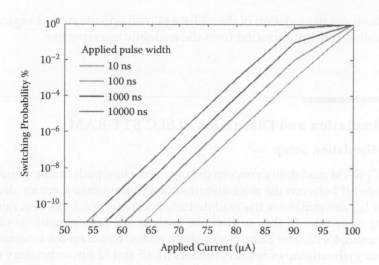

FIGURE 3.4
Switching probability with 200 mA critical current.

we shall show in the following sections. We note that material parameter fluctuations may cause the variations of the switching probability of MTJ by introducing the normally distributed localized fluctuation of magnetic anisotropy, etc.

3.3.2 Reading Sensing Margin

Following the scaling of switching current, the read current of the MTJ must decrease accordingly to suppress the read disturbance. This fact will introduce another concern on the readability of STT-RAM: in a traditional current sensing STT-RAM read scheme, for instance, a read current I_{read} is injected into the memory cell. The generated bit-line voltage is then compared to a reference voltage to read out the MTJ resistance state. The generated sense margin, which can be measured by the voltage difference between the bit-line voltage and the reference voltage, is proportional to $I_{read} \cdot R_L \cdot TMR$. Here R_L is the low MTJ resistance. TMR is tunnel magnetoresistance ratio. A certain sense margin must be maintained in STT-RAM read operations to overcome the device mismatch in the sense amplifier and keep the sensing errors at a minimum level.

When I_{read} decreases, the generated sense margin of STT-RAM may reduce if the MTJ resistance or TMR does not increase proportionately. The degraded sense margin may incur sensing errors if the device variation of the sense amplifier is large. Because the process variations of CMOS technology become more and more severe when manufacturing technology scales, readability may replace the write failure as the limiting factor of the reliability of the STT-RAM design. It is necessary to conduct a detailed analysis on the robustness degradation of the STT-RAM read operations and explore the MTJ scaling rule optimization from the readability perspective.

3.4 Simulation and Discussion of SLC STT-RAM

3.4.1 Simulation Setup

The STT-RAM readability concern during the scaling path can be considered the trade-off between the read disturbance and the sensing errors: decreasing the I_{read} can minimize the read disturbance; however, it can also raise the sensing error rate. In this section, we analyze the read operations of STT-RAM under 45, 32, and 22 nm technology nodes. Based on the International Technology Roadmap of Semiconductors [1], 45 and 32 nm technology nodes will still use in-plane MTJ, while perpendicular MTJ will be adopted in 22 nm technology.

Both read disturbance and sensing errors are included in our experiments: we use the stochastic LLG equation to simulate the magnetization switching of the MTJ. Thermal fluctuation is represented by a normalized thermal agitation fluctuating field [23]. We also conduct Monte-Carlo simulations to analyze the impacts of process variations on the sense amplifiers' sensing margin and the sensing errors of STT-RAM. Table 3.1 shows the nominal STT-RAM cell parameters at different technology nodes, which are extracted from [5]. Following the scaling of MTJ and NMOS transistor devices, the resistance states of the MTJ increase accordingly. The predictive technology model (PTM) is used in our NMOS transistor simulation [4].

3.4.2 Reading Operation Error Analysis

As mentioned in Section 3.3, two major error sources of STT-RAM read operations are (1) sensing errors and (2) read disturbance. Here the sensing

TABLE 3.1

Summary of STT-RAM Parameters

Technology (nm)	MTJ Geometry (nm)			R_H (Ω)	R_L (Ω)	Transistor Size (nm)
	Length	Width	Oxide Thickness			
45	90	40	2.2	2000	1000	270
32	65	32	2.2	2500	1000	150
22	45	22	3.2	4000	1500	75

Source: Y. Chen et al., *IEEE Transactions on Very Large Scale Integration (VLSI) Systems*, 18(12):1724–1734, 2010. With permission.

error denotes the case that the readout voltage difference between the bit-line voltage and reference voltage is smaller than the required sense margin. In such a case, the MTJ resistance state could not be read out correctly, for example, reading out as a 1 when storing a 0, or reading out as a 0 when storing a 1. In an STT-RAM read scheme, we normally apply I_{read} only from one direction. Due to the asymmetry of MTJ switching, the MTJ flipping from 0 to 1 is more difficult than the other direction. In our simulation, we assume the I_{read} is applied from the direction that ensures the read disturbance happening only when the cell stores a 0 (and may possibly flip to 1). In the rest of this section, we discuss these two error sources separately.

3.4.2.1 Sensing Error

For a fair comparison, the same sense amplifier structure is used in the simulations under different technology nodes. The ratio between the channel widths of the NMOS and PMOS transistors is maintained the same while the channel length of the transistors is adjusted according to the technology node. We define the sense margin as the voltage difference actually generated on the two inputs of the sense amplifier. A high sensing margin generally corresponds to a low sensing error rate. Because of the increased process variations, the sense margin of the sense amplifier must increase to overcome the device mismatch in the sense amplifier as the transistor feature size reduces. Figure 3.5(a) shows our simulation result of the sense margins at various technologies. When technology scales from 45 nm to 22 nm, the read currents (20% of I_{C0}) decrease from 34 µA to 15 µA. The nominal sense margins generated from the STT-RAM cell are 17, 18, and 18.8 mV, respectively. Thanks to the improvement of resistance and TMR due to the adoption of perpendicular MTJ, the sense margin even slightly increases when the technology scales.

The sensing errors occur when the voltage difference on the inputs of the sense amplifier cannot overcome the device mismatch of the circuit. Figure 3.5(b) shows the sensing error rates at different technology nodes when I_{read} changes. The device variations of both MTJ and NMOS transistors are included in our simulation. The standard deviation of geometry size is set to 5% of the nominal value, while the standard deviation of the NMOS transistor threshold voltage is set to 30 mV. Following the increase of I_{read}, the sensing error rate reduces rapidly. The highest sensing error rate always occurs at 22 nm for the same reading current ratio I_{read}/I_{C0}, due to the combined impacts of the small critical switching current and the large variations of MTJ and CMOS devices.

We note that in our sense amplifier designs at different technologies, the transistor sizes follow a simple scaling rule, i.e., multiplied by the inverse of the technology scaling factor. Increasing supply voltage (V_{dd}) can improve the robustness of the sense amplifier and reduce the sense margin. In our simulation, we set V_{dd} to 1.0 V for the minimization of the leakage power.

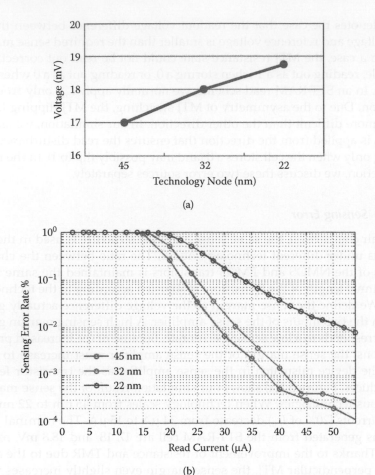

(a)

(b)

FIGURE 3.5

(a) Sensing margin trend of STT-RAM sensing circuit. (b) Sensing error rate versus reading current.

3.4.2.2 Reading Disturbance

Normally the read current is set to about 20% of the write currents to minimize the read disturbance. For example, the read currents of 34, 24, and 15 μA are usually selected 45, 32, and 22 nm technology nodes, respectively. However, as shown in Figure 3.5, such low read currents could cause a severe sensing error rate in the scaled technologies. Increasing the read current, however, will raise the reading disturbance [23].

The simulated STT-RAM cell read disturbance rates under different I_{read}/I_{C0} ratios are shown in Figure 3.6. In all test cases, the read disturbance quickly increases when the I_{read} rises.

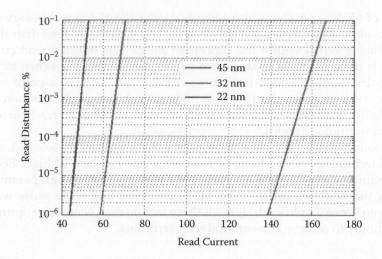

FIGURE 3.6
Reading disturbance versus reading current.

3.4.3 Optimize Discussion

Because the trends of STT-RAM read disturbance and sensing errors are opposite when the I_{read} changes, it is possible for us to find the optimal I_{read} that can achieve the minimum total read operation error rate.

Figures 3.7(a)–(c) show our combined simulated results of the STT-RAM read operation error rate under different read currents at the technology

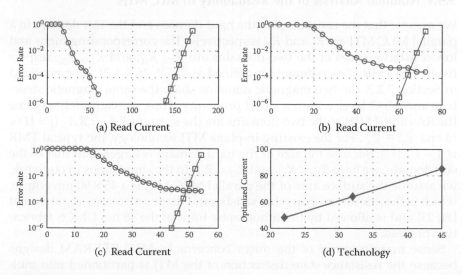

FIGURE 3.7
Reading disturbance versus reading current.

nodes of 45, 32, and 22 nm, respectively. Under the 45 nm technology node, there is almost no intersection between the trend curves of read disturbance and sensing errors within the concerned range. The optimal read current, which is the cross-point of two error curves, is about 80 μA. When technology scales to lower nodes, both error curves shift. The cross-point of two error curves increases, leading to a higher combined read operation error rate. Figure 3.7(d) shows the simulated optimal point of the read current at various technology nodes.

We found that even when the optimal read current is selected at the 22 nm technology node, the total read error rate can be still high. Because the optimal read current is also affected by many other design parameters, that is, the NMOS transistor and MTJ sizes, read and write pulse widths, and even sense amplifier design, all these factors may need to be optimized accordingly to achieve a lower total read error rate.

3.5 Readability Analysis of MLC MTJs

The read and write operations of MLC STT-RAM cells are much more complicated than those of SLC STT-RAM cells. In this section, we mainly discuss the readability of MLC MTJs. The writability of MLC MTJs will be investigated thereafter.

3.5.1 Nominal Analysis of the Readability of MLC MTJs

We assume that the resistances of the hard domain and the soft domain in a parallel MLC MTJ are R_1 and R_2, respectively. The corresponding high- and low-resistance states of the two domains are R_{1H}, R_{1L}, and R_{2H}, R_{2L}, respectively. The TMR of each domain is defined as $\frac{R_{iH} - R_{iL}}{R_{iL}}$, $(i = 1, 2)$. As mentioned in Section 3.2.3, the two magnetic domains share the same magnetic structure and MgO layer within a small proximity. Thus, we can safely assume the RAs and TMRs of the two domains are the same, or $RA_{1j} = RA_{2j}$ ($j = H$ or L) and $\frac{R_{2H}}{R_{1L}} = \frac{R_{2H}}{R_{2L}}$. For the existing in-plane MTJ technology, the typical TMR is 1 ~ 1.2 [13]. Because the size of the hard domain is larger than that of the soft domain, we have $R_{1H} < R_{2H}$ and $R_{1L} < R_{2L}$. In the simulations in our work, we assume the surface area of the parallel MLC MTJ is a 45 × 90 nm ellipse, which is the minimum shape that satisfies the shape anisotropy requirement [11, 23] and is allowed by the lithography limit of the 45 nm CMOS fabrication process.

Sense margin is one of the major concerns in MLC STT-RAM designs because the resistance state distinction of the MTJ is partitioned into multiple levels. Read errors happen when the distributions of the two adjacent resistance states (i.e., 00 vs. 01, 01 vs. 10, and 10 vs. 11) overlap with each other,

TABLE 3.2

Distinction between the Adjacent Resistance States
of Parallel and Series MLC MTJs

State Distinctions	Parallel MLC MTJ
$R_{01} - R_{00}$	$\frac{TMR \cdot R_{1L}^2 R_{2L}}{(R_{1L}+R_{2L})(R_{1L}+(TMR+1)R_{2L})}$
$R_{10} - R_{01}$	$\frac{TMR \cdot (TMR+1) \cdot R_{1L} R_{2L} (R_{2L}-R_{1L})}{(R_{1L}+(TMR+1)R_{2L})((TMR+1)R_{1L}+R_{2L})}$
$R_{11} - R_{10}$	$\frac{TMR \cdot (TMR+1) \cdot R_{1L}^2 R_{2L}}{(R_{1L}+R_{2L})((TMR+1)R_{1L}+R_{2L})}$
	Series MLC MTJ
$R_{01} - R_{00}$	$TMR \cdot R_{1L}$
$R_{10} - R_{01}$	$TMR(R_{2L} - R_{1L})$
$R_{11} - R_{10}$	$TMR \cdot R_{1L}$

or the distinction between the two resistance states is smaller than the sense
amplifier resolution. The reading error rate can be reduced by maximiz-
ing the distinctions between every two adjacent states, which are shown in
Table 3.2. Without considering the process variations, the goal of the nominal
design method of the MLC STT-RAM cell is to maximize the distinctions
between every two adjacent resistance states.

In the real implementation of parallel MLC MTJs, $R_{00} = R_{1L}||R_{2L}$ and
$R_{11} = R_{1H}||R_{2H}$ are fixed by the MTJ designs. The changes of R_{01} and R_{10} are
not independent and determined by the partitioning of the free layer. If we
assume the surface area of the parallel MLC MTJ is A and the surface area of
the hard domain is A_1, we have

$$R_{1L} A_1 = R_{2L} \cdot (A - A_1) = R_{00} \cdot A \tag{3.4}$$

$$R_{1H} A_1 = R_{2H} \cdot (A - A_1) = R_{11} A \tag{3.5}$$

Here $A_1 > A/2$. The distinctions between every two adjacent resistance states
can be calculated as

$$D_{00-01} = R_{01} - R_{00} = \frac{TMR \cdot RA}{A} \cdot \frac{A - A_1}{A + A_1 \cdot TMR}, \tag{3.6}$$

$$D_{00-01} = R_{10} - R_{01} = \frac{[TMR \cdot (TMR+1) \cdot RA](2A_1 - A)}{(A + TMR \cdot A_1)[TMR \cdot (A - A_1) + A]}, \tag{3.7}$$

$$D_{10-11} = R_{11} - R_{10}$$

$$= \frac{TMR \cdot (TMR+1) \cdot RA}{A} \cdot \frac{A - A_1}{TMR \cdot (A - A_1) + A}. \tag{3.8}$$

We calculated the derivatives of D_{00-01}, D_{01-10}, and D_{10-11} with respect to A_1 and have $\frac{dD_{00-01}}{dA_1} < 0$, $\frac{dD_{10-01}}{dA_1} < 0$, and $\frac{dD_{01-10}}{dA_1} > 0$ when $A_1 \in [A/2, A]$. In other words, D_{00-01} and D_{10-11} monotonically decrease when A_1 increases from $A/2$ to A, and D_{01-10} monotonically increases in the same range. Also, because $A - A_1 < A_1$ and $TMR \geq 1$, D_{10-11} is always larger than D_{00-01} based on Equations (3.6) and (3.8). Therefore, the optimal design of parallel MLC MTJs happens when $D_{00-01} = D_{01-10}$, or

$$(TMR+1)(\tfrac{R_{2L}}{R_{1L}})^2 - \tfrac{R_{2L}}{R_{1L}} = 2(TMR+1). \tag{3.9}$$

Here $R_{1L} || R_{2L} = R_{00}$.

In a series MLC MTJ, the optimal MTJ design happens when $D_{00-01} = D_{01-10} = D_{10-11}$, or

$$R_{1L} = \tfrac{1}{2} R_{2L}. \tag{3.10}$$

Here R_{2L} is usually the low-resistance state of the SLC MTJ with the minimum surface area (say, A). The optimal design parameters of a typical parallel MLC MTJ and a typical series MLC MTJ are shown in Table 3.3.

3.5.2 Statistical Analysis of the Readability of MLC MTJs

All the optimizations in Section 3.5.1 are based on the nominal values of the device parameters of MLC MTJs. In this section, we will analyze the impacts of process variations on the readability of MLC STT-RAM cells.

Figures 3.8(a) and (b) show the distributions of the four resistance states in a parallel MLC MTJ and a series MLC MTJ, respectively. Both MTJs are optimized by using the nominal optimization method presented in Section 3.5.1.

TABLE 3.3

Optimal Design Parameters

Parameter	Mean	1σ Deviation
RA ($\Omega\mu m^2$)	20	7%
Oxide thickness (nm)	2.2	2%
TMR	1.2	9%
Surface area	Width × length (nm^2)	
Limitation size	45 × 90	5%
Parallel MLC MTJ	45 × 90	5%
R_1/R_2 in parallel MLC MTJ	1.66 (nominal)	N/A
R_1/R_2 in parallel MLC MTJ	2.2 (statistical)	N/A
R_1 in series MLC MTJ	45 × 90	5%
R_2 in series MLC MTJ	64 × 127 (nominal)	5%
R_2 in series MLC MTJ	64.5 × 129 (statistical)	5%

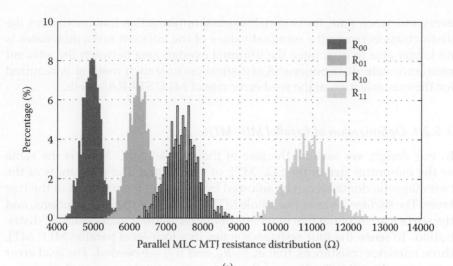

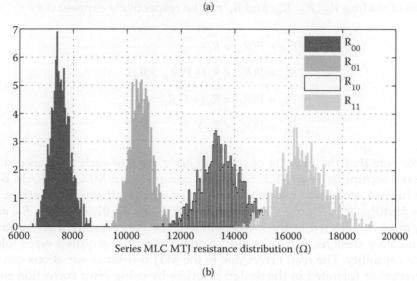

FIGURE 3.8
Four state resistance distributions of (a) parallel MLC MTJ and (b) series MLC MTJ, optimized by nominal design method.

The standard deviations (1σ) of RA and TMR are 7 and 9%, respectively, based on the measurement data in [13]. All other parameters used in our simulations are summarized in Table 3.3. In the nominal optimized parallel MLC MTJ, $\frac{R_1}{R_2} = 1.66$. In the nominal optimized series MLC MTJ, the surface area of the larger MTJ is 64×127 nm, which corresponds to a low-resistance state of $R_{2L} = 2500 \ \Omega$. After the process variations are taken into account, the distributions of the resistance states overlap with each other, resulting in the read errors of the MLC MTJ. Because of the different deviations of every resistance state, the original nominal optimization that maximizes the distinctions between the nominal values of the adjacent resistance states is no longer able to guarantee the minimal overlap area between the adjacent resistance state distributions. A statistical optimization method is required for the minimization of the read error rate of MLC STT-RAM cells.

3.5.2.1 Optimization of Parallel MLC MTJs

In our design, we assume the size of the parallel MLC MTJs is the same as the minimum size of the SLC MTJ, or 45×90 nm. The resistances of the two magnetic domains can be adjusted by changing the partition of the free layer. The surface areas of the whole MTJ follow Gaussian distributions, and the surface areas of the two magnetic domains follow a joint Gaussian distribution. To sense the four resistance states in a four-level parallel MLC MTJ, three reference resistances, that is, R_I, R_{II}, and R_{III}, are needed. The read error rates of reading R_{00}, R_{01}, R_{10}, and R_{11} can be respectively expressed as

$$
\begin{aligned}
P_{e00} &= P(R_{00} > R_I) \\
P_{e01} &= P(R_{01} < R_I) + P(R_{01} > R_{II}) \\
P_{e10} &= P(R_{10} < R_{II}) + P(R_{10} > R_{III}) \\
P_{e11} &= P(R_{11} < R_{III})
\end{aligned}
\tag{3.11}
$$

We note that the impacts of the read error rates of each resistance state are not accumulative in MLC STT-RAM designs: for a MLC STT-RAM cell, the highest read error rate is the maximum one of all resistance states, or $P_e = Max(P_{e00}, P_{e01}, P_{e10}, P_{e11})$. To minimize the P_{ei} ($i = 00, 01, 10, 11$), R_I, R_{II}, and R_{III} must be selected at the cross-points of the two adjacent distributions. In memory designs, P_e can be used to determine the required error tolerance capability. The read errors due to the MTJ resistance variations can be corrected or tolerated in the design practices by using error correction code (ECC), design redundancy, etc.

In Figure 3.8(a), the overlaps of the resistance state distributions of the parallel MLC MTJ generate the read error rates of $P_{e00} = 0.73\%$, $P_{e01} = 6.44\%$, $P_{e10} = 6.05\%$, and $P_{e11} = 0.018\%$. High read error rates happen at R_{00} and R_{01}, which are incurred by the large overlaps between these two resistance states.

If we assume that the low resistance of each magnetic domain follows a Gaussian distribution as $R_{iL} \sim N(R_{iL}, \sigma_i)(i = 1, 2)$, the distribution of $R_{iL}(i = 1, 2)$ can be expressed as

$$
f_{iL}(x_i) = \frac{1}{\sqrt{2\pi}\sigma_i} \frac{RA}{x_i^2} e^{-\frac{(RA/x_i - R_{iL})^2}{2\sigma_i^2}}.
\tag{3.12}
$$

Ishigaki et al. [13] show that the TMR also follows a Gaussian distribution. We introduce a new variable $z = TMR + 1$. In our simulations, $z \sim N(2.2, 9\% \times 1.2)$ and its distribution can be expressed as

$$f_0(x_i) = \frac{1}{0.108\sqrt{2\pi}} e^{\frac{(z-2.2)^2}{0.023328}} \cdot R_{iH} = z \cdot R_{iL}.$$

Then the read error rate probability of every resistance state of the parallel MLC MTJ can be further derived from Equation (3.11) as

$$P(R_{00}) = \frac{1}{\sqrt{2\pi}(\frac{\sigma_1+\sigma_2}{RA})} \frac{1}{R_{00}^2} \cdot e^{\frac{(1/R_{00}-(R_{1L}+R_{2L})/RA)^2}{2(\frac{\sigma_1+\sigma_2}{RA})^2}}$$

$$P(R_{01}) = \int_{-\infty}^{+\infty} f_{2L}(\tfrac{RA}{R_{01}} - x) \int_{-\infty}^{+\infty} f_{1L}(z) f_0\left(\tfrac{z}{x}\right) \tfrac{z}{x^2} \, dz dx \tfrac{RA}{R_{01}^2}$$

$$P(R_{10}) = \int_{-\infty}^{+\infty} f_{1L}(\tfrac{RA}{R_{10}} - x) \int_{-\infty}^{+\infty} f_{2L}(z) f_0\left(\tfrac{z}{x}\right) \tfrac{z}{x^2} \, dz dx \tfrac{RA}{R_{10}^2}$$

$$P(R_{11}) = \int_{-\infty}^{+\infty} \int_{-\infty}^{+\infty} f_{1L}(x) f_0\left(\tfrac{x}{y}\right)\left(\tfrac{x}{y^2}\right) dx$$

$$\cdot \int_{-\infty}^{+\infty} f_{2L}(x') f_0(\tfrac{x'}{R_{11}-y}) \tfrac{x'}{(R_{11}-y)^2} dx' dy \qquad (3.13)$$

Figure 3.9 depicts the Monte-Carlo simulation results of the read error rate under the different ratios of the nominal resistances of the two magnetic

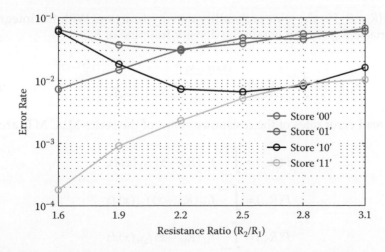

FIGURE 3.9
Error rate versus R2/R1 ratio sweep.

domains (R_2/R_1) when the MTJ variations are considered. P_{e11} is always lower than P_{e00} due to the bigger distinction between R_{10} and R_{11} compared to the one between R_{00} and R_{01}. Following the increase of R_2/R_1 from 1.6, both P_{e00} and P_{e11} increase, indicating the reduced distinction from the adjacent resistance states. However, the increase of R_2/R_1 decreases the P_{e01} and P_{e10} by raising the distinction between R_{01} and R_{10}.

When $R_2/R_1 = 2.2$, the parallel MLC MTJ achieves its lowest maximum read error rate as $P_{e00} = 3.31\%$, $P_{e01} = 2.97\%$, $P_{e10} = 0.73\%$, and $P_{e11} = 0.23\%$. The change of the optimal R_2/R_1 ratios in the nominal and statistical optimizations comes from the correlation between the standard deviation and the nominal values of the MTJ resistance state: the higher the resistance is, the larger the standard deviation of the resistance will be [24].

3.5.2.2 Optimization of Series MLC MTJs

In a series MLC MTJ, the serially connected SLC MTJs are fabricated separately. The parameters of these two MTJs are partially correlated due to the spatial correlations. The two resistance states of the small SLC MTJ with the minimum size are $R_{2L} = 5000\ \Omega$ and $R_{2H} = 11,000\ \Omega$, respectively. The distinctions between two adjacent resistance states can be adjusted by changing the surface area of the large SLC MTJ.

Similar to the analysis in Section 3.5.2, we have

$$z = R_{iH}/R_{iL} \tag{3.14}$$

and

$$g(R_{iH}, y_i) = f_{iL}(R_{iL}) \cdot f_0(\tfrac{R_{iH}}{R_{iL}}) \cdot \tfrac{1}{R_{iL}}. \tag{3.15}$$

Here $g(R_{iH}, R_{iL})$ is the joint probability density function. With the integral of the variable R_{iL}, we can obtain the density function of R_{iH}.

$$f_{iH}(x_i) = \int_{-\infty}^{+\infty} \frac{1}{\sqrt{2\pi}\sigma_i} \frac{RA}{R_{iL}^2} e^{\frac{(RA/R_{iL}-R_{iL})^2}{2\sigma_i^2}} \cdot \frac{1}{R_{iL}} \cdot \frac{1}{0.108\sqrt{2\pi}} e^{\frac{(\frac{R_{iH}}{R_{iL}}-1.2)^2}{0.023328}} dR_{iL}. \tag{3.16}$$

The read error rates of the resistance states of the series MLC MTJ are

$$P(R_{00}) = \int_{-\infty}^{+\infty} f_{1L}(R_{00} - x) f_{2L}(x) dx$$

$$P(R_{01}) = \int_{-\infty}^{+\infty} f_{1H}(R_{01} - x) f_{2L}(x) dx$$

$$P(R_{10}) = \int_{-\infty}^{+\infty} f_{1L}(R_{10} - x) f_{2H}(x) dx$$

$$P(R_{11}) = \int_{-\infty}^{+\infty} f_{1H}(R_{11} - x) f_{2H}(x) dx \tag{3.17}$$

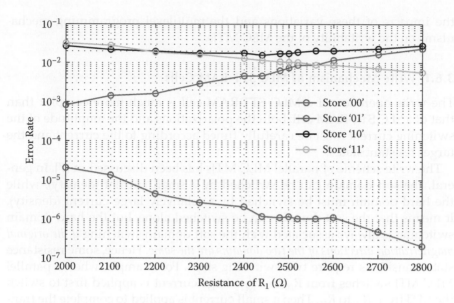

FIGURE 3.10
Error rate versus resistance of hard domain sweep.

Figure 3.10 shows the read error rates of the four resistance states of the series MLC MTJ when sizes of the large SLC MTJ changes. The variation of the large SLC MTJ size is represented by its low-resistance state (R_{1L}). The lowest maximum read error rate happens when $R_{1L} = 2440$ Ω, or the MTJ size is 64.5×129 nm. It is very close to the result of the nominal optimization method, $R_{1L} = 2500$ Ω, or the MTJ size of 64×127 nm. The corresponding read error rates of each resistance state are $P_{e00} = 0.000118\%$, $P_{e01} = 0.46\%$, $P_{e10} = 1.57\%$, and $P_{e11} = 1.15\%$. Compared to parallel MLC MTJs, series MLC MTJs demonstrated a significantly lower read error rate under the same fabrication conditions. Although the read error rate has not achieved the commercial requirement yet, these results are still very encouraging.

3.6 Writability Analysis of MLC MTJs

In SLC MTJ designs, increasing the switching current density can effectively reduce the MTJ switching time and improve the write error rate of the SLC STT-RAM cell. In MLC MTJ designs, however, increasing the switching current when programming the MTJ to an intermediate resistance state may overwrite the MTJ to the next resistance level. The thermal fluctuations further complicate the situations of MLC MTJ programming by incurring the additional variability of MTJ switcing time. In this section, we will discuss the impacts of these variations and the multilevel programming mechanisms on the writability of the MLC MTJs.

3.6.1 Write Mechanism of MLC STT-RAM Cells

The write operation of a MLC STT-RAM cell is much more complex than that of a SLC STT-RAM cell: both the polarizations and the amplitude of the switching current must be carefully tuned according to the current and the target resistant states.

The write scheme of parallel MLC MTJs has been discussed in [7]. In general, the soft domain can be switched by a small current (density), while the hard domain must be switched by a relatively large current (density). It means that the soft domain can be switched alone, but the hard domain switching is always associated with the soft domain switching *if the original magnetization directions of the two domains are the same*. Hence, some resistance state transitions require two switching steps. For example, when a parallel MLC MTJ switches from R_{00} to R_{10}, a large current is applied first to switch the MTJ from R_{00} to R_{11}. Then a small current is applied to complete the transition from R_{11} to R_{10}.

The write mechanism of series MLC MTJs is slightly different from that of parallel MLC MTJs. The switching of the large (small) SLC MTJ requires a high (low) switching current, denoting the change of the lower (higher) bit of the two-bit data. The switching of the large SLC MTJ is always associated with the small SLC MTJ switching if the original magnetization directions of the two MTJs are the same. Figure 3.11 summarizes the transition graphs of the write mechanisms of the two types of MLC MTJs. As summarized in [6], the transitions of the MTJ resistance states can be classified into three types:

1. Soft transition (ST), which switches only the soft domain in a parallel MLC MTJ or the small SLC MTJ in a series MLC MTJ

2. Hard transition (HT), which switches both domains in a parallel MLC MTJ or both SLC MTJs in a series MLC MTJ to the same magnetization direction

3. Two-step transition (TT), which utilizes two steps to switch the MLC MTJ to the target resistance states, i.e., one HT followed by one ST.

3.6.2 Impacts of Thermal Fluctuations

We define the threshold switching current (density) as the minimal current density required to switch a MTJ within a switching time. The relationship between the magnetization switching time (t_w) and the nominal value of the threshold switching current density (J_C) can be expressed by three equations based on the working regions [21]:

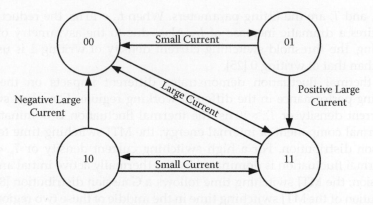

(a) State transition of writing operation
in parallel MLC STT-RAM

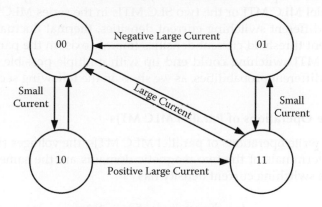

(b) State transition of writing operation
in series MLC STT-RAM

FIGURE 3.11
Write transition mechanisms of two types of MLC MTJs. (a) Parallel MLC MTJ. (b) Series MLC MTJ.

1. $J_{C1}(t_w) = J_{C0}\left(1 - (1/\Delta)\ln(t_w/\tau_0)\right)$, $(t_w > 10ns)$;

2. $J_{C2}(t_w) = \frac{J_{C1}(t_w) + J_{C3}(t_w)\exp(-B(t_w-T))}{1+\exp(-A(t_w-T))}$, $(3ns < t_w < 10ns)$;

3. $J_{C3}(t_w) = J_{C0} + C\ln(\pi/2\theta)/t_w$, $(t_w < 3ns)$;

Here J_{Ci} ($i = 1, 2, 3$) are the threshold switching current densities in the three working regions, respectively, J_{C0} is the threshold switching current density that causes a spin flip at 0 K without any external magnetic fields, τ_0 is relaxation time, T is the working temperature, and θ is the initial angle between the magnetization vector and the easy axis. The remaining parameters, such as C, B, and T, are the fitting parameters. When $t_w < 10$ ns, the reduction of t_w requires a dramatic increase of J_C. Also, due to the asymmetry of MTJ switching, the threshold switching current density of writing 1 is usually larger than that of writing 0 [25].

The thermal fluctuation demonstrates different impacts on the MTJ switching performance in the different working regions: For a low switching current density or $T_w > 10$ ns, the thermal fluctuation is dominated by the thermal component of internal energy; the MTJ switching time follows a Poisson distribution. For a high switching current density or $T_w < 3$ ns, the thermal fluctuation is dominated by the thermally active initial angle of procession; the MTJ switching time follows a Gaussian distribution [8]. The distribution of the MTJ switching time in the middle of these two regions follows a combination of the two distributions. In the write operations of MLC STT-RAM, the two parts of the MLC MTJs, that is, the two magnetic domains in the parallel MLC MTJ or the two SLC MTJs in the series MLC MTJ, may experience different switching current densities, thermal fluctuations, and even different threshold current densities (mainly exist in the parallel MLC MTJs). The MTJ switching could end up with multiple possible resistance states with different probabilities, as we show in the following sections.

3.6.3 Write Operations of Parallel MLC MTJs

During the write operations of parallel MLC MTJs, the voltages (V) applied to the two terminals of the two magnetic domains are the same. For each domain, the switching current density has

$$J_i = \frac{V}{R_i \cdot A_i} = \frac{V}{\frac{RA_i}{A_i} \cdot A_i} = \frac{V}{RA_i}, i = 1, 2. \tag{3.18}$$

This shows that after V is fixed, the switching current density through each domain is uniquely determined by the RA of the domain. Here $RA_i = RA_L$ or $RA_L \cdot (TMR + 1)$ for the low- or the high-resistance state, respectively. RA_L is the RA of the low-resistance state. As we discussed in Section 3.2.3, the two magnetic domains of a parallel MLC MTJ have exactly the same RA when they are in the same resistance state. In such a case, the two magnetic domains have the same current density. However, if the two domains are in opposite resistance states, their current densities will be different.

Figure 3.12(a) shows our simulated results of the relationships between the T_w and J_C for the two domains in a typical parallel MTJ. The MTJ parameters are scaled from the measured data of a 90×180 nm elliptical MTJ device in [17]. Two domains demonstrate different J_C even under the same T_w due to the different shape anisotropies, etc. The write asymmetry is also observed in the result; that is, the J_C of the $0 \rightarrow 1$ transition of the magnetic domain is always higher than that of the $1 \rightarrow 0$ transition for the same T_w. The relative

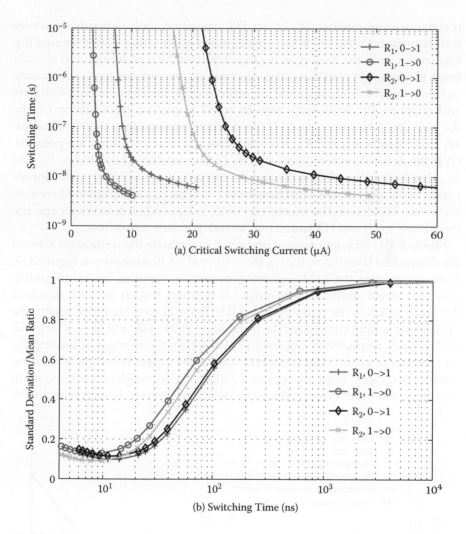

FIGURE 3.12
Switching properties of the two domains for a parallel MLC MTJ. (a) Switching time versus switching current. (b) Switching time standard deviation versus switching current.

deviations of the T_w of the two magnetic domains at the whole working region are shown in Figure 3.12(b).

During the write operations of parallel MLC STT-RAM cells, the write current must be applied to switch only the domain(s) that need(s) to be flipped. However, the variability in the magnetization switching of the two domains can introduce write errors. Different from the SLC MTJ where the write error is only incurred by incomplete switching, the writing errors of the parallel MLC MTJ come from either the incomplete switching of the target domains (incomplete write) or overwriting the other domain to an undesired resistance state (overwrite). In an HT transition, only incomplete writes will happen because the write operations require either both domains flip together or only the hard domain flips because the soft domain has already been in the target resistance state. In such a case, increasing the switching current can effectively improve the switching performance of both domains and suppress the write error rate. In an ST transition, the situation can be divided into two scenarios: (1) If the destination resistance state is a boundary state, that is, R_{00} and R_{11}, the only incomplete write failures are possible. (2) If the destination resistance state is an intermediate state, that is, R_{01} and R_{10}, then both incomplete write and overwrite failures may occur. An appropriate switching current must be selected to achieve a low combined writing error rate. We denote the transitions in (2) as dependent transitions and the transitions in (1) and HT transitions as independent transitions.

Monte-Carlo simulations are conducted to evaluate the write error rates of the dependent transitions, that is, $00 \rightarrow 01$ or $11 \rightarrow 10$, as shown in Figure 3.13. Here we assume the MTJ switching current is supplied by an adjustable on-chip current source, whose output magnitude has an intrinsic standard deviation of 2% of the nominal value [15]. For a 10 ns write pulse width, the optimal switching current for the transitions of $00 \rightarrow 01$ and $11 \rightarrow 10$ are 46.5 and 49.9 µA, respectively. Figure 3.13 also shows the changes of incomplete and overwrite errors over the whole simulated range. When the switching current decreases from the optimal value, the incomplete writes start to dominate the write errors; when the switching current increases from the

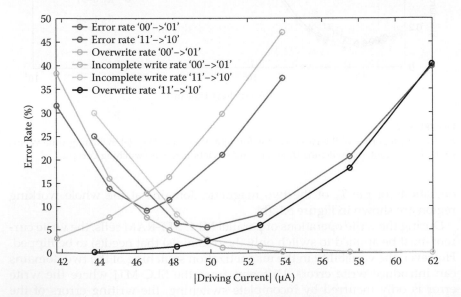

FIGURE 3.13
See color insert. Writing error rate in parallel MLC STT-RAM cell at $T_w = 10$ ns.

optimal value, the overwrite errors of the hard domain start to dominate the write errors. Nonetheless, the error rates of the two dependent transitions are still high (8.2%), indicating a large overlap area between the threshold switching current distributions of the hard domain and the soft domain.

Figure 3.14 shows the write error rates of the dependent transitions of the parallel MLC MTJ at different switching currents when $T_w = 3$, 10, and 100 ns, respectively. The lowest write error rate is achieved at $T_w = 3$ ns. This is because when T_w reduces, the required MTJ switching current increases. The impact of the thermal fluctuations on the MTJ switching is suppressed and the distributions of the T_w are compressed. This fact indicates that the parallel MLC MTJ better work at a fast working region to minimize the write error rate.

We can also map the uncertainties in the switching time of the parallel MLC MTJ under the fixed switching currents into the distributions of the required switching currents for fixed switching times. Figure 3.15(a) shows the distributions of the threshold switching current of the dependent transitions for the parallel MLC MTJ at a 10 ns write pulse width. The distributions of the MTJ write currents supplied by the on-chip current source are also depicted. Take the transition of $00 \rightarrow 11$ as an example; a write current is selected between the threshold current distributions of the transitions of $00 \rightarrow 01$ and $00 \rightarrow 11$. The two types of write errors, including incomplete write and overwrite, are represented by the overlap between the distributions of the write current and the threshold switching current of $00 \rightarrow 01$ and the overlap between the distributions of the write current and the threshold switching current of $00 \rightarrow 11$, respectively. Figure 3.15(b) shows the distributions of the threshold switching current of the independent transitions for the parallel MLC MTJ at a 10 ns write pulse width. Because only the target

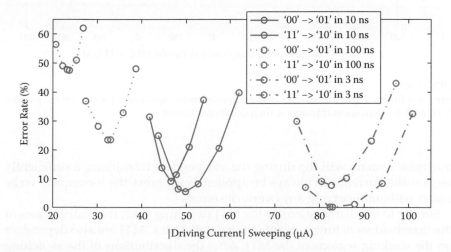

FIGURE 3.14
Writing error rate in a parallel MLC STT-RAM cell at different T_w values.

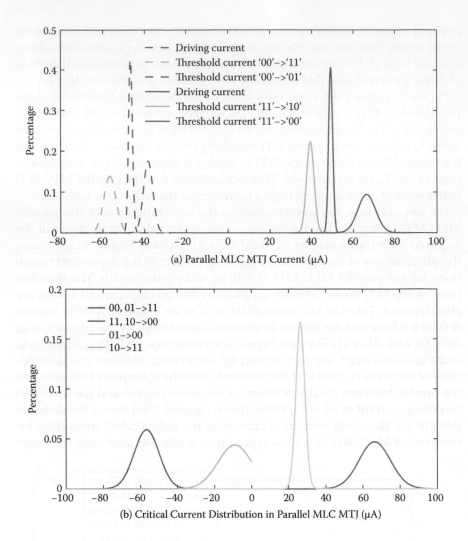

(a) Parallel MLC MTJ Current (μA)

(b) Critical Current Distribution in Parallel MLC MTJ (μA)

FIGURE 3.15
See color insert. Threshold current distributions of resistance state trasitions for the parallel MLC MTJ. (a) Dependent transitions. (b) Independent transitions.

magnetic domain will flip during the independent transitions, a sufficiently large write current can always be applied to suppress the incomplete write errors without incurring any overwrite errors.

Similar to the distributions of the MTJ switching time, the distributions of the threshold switching current of the parallel MLC MTJ are also dependent on the working regions of the MTJ. After the distributions of the switching current of the resistance state transitions are obtained, the optimal write current can be also derived as in Figure 3.15(a).

3.6.4 Write Operations of Series MLC MTJs

In a series MLC MTJ, the magnitudes of the currents passing through the two SLC MTJs are the same. However, the applied current densities on the two SLC MTJs are different and determined by the different surface areas of them. In Section 3.5.2, the analysis on the read reliability of the series MLC MTJs shows that the optimal surface area ratio between the two MLC MTJs is around 2, or 45×90 nm and 64.5×129 nm at the 45 nm technology node. In our simulations, we also assume the two SLC MTJs maintain the same aspect ratios and were fabricated under the same conditions. Thus, they have the same switching properties, that is, the same relationships between threshold switching current density and the switching time. Again, the switching current density on each SLC MTJ is controlled by the on-chip write current source.

Figure 3.16 shows the write error rates of the dependent transitions of the series MLC MTJ under different switching currents for a 10 ns write pulse width. The optimal switching currents for the transitions of $00 \rightarrow 10$ and $11 \rightarrow 01$ are 79.0 and 92.5 μA, respectively. Compared to parallel MLC MTJs, the write error rates of the dependent transitions are significantly reduced: the minimum write error rates of the transitions of $00 \rightarrow 10$ and $11 \rightarrow 01$ are only 0.0015 and 0.0043%, respectively. The improvement in the write reliability is because of the larger distinction between the threshold switching current distributions of the dependent transition and the adjacent resistance state transition, as shown in Figure 3.17(a). For comparison purposes, the results of the independent resistance state transitions are shown in Figure 3.17(b).

Figure 3.16 also shows the write error rates of the dependent transitions of the series MLC MTJ at different switching currents when $T_w = 3$ and 100 ns,

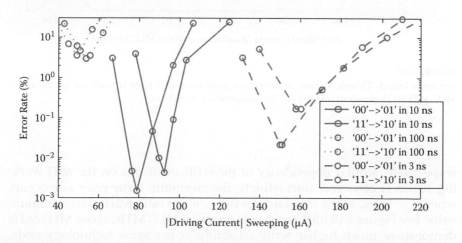

FIGURE 3.16
Writing error rate in a series MLC STT-RAM cell at different T_w values.

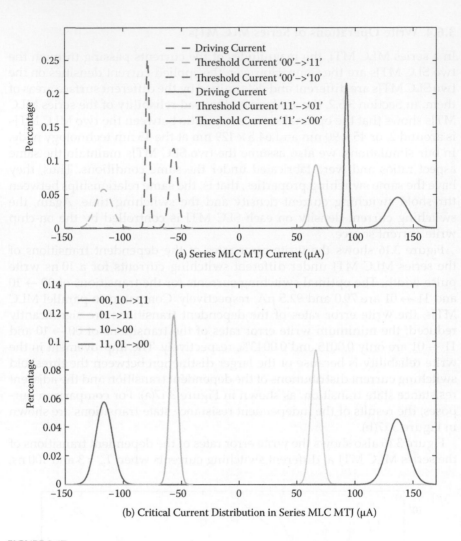

(a) Series MLC MTJ Current (μA)

(b) Critical Current Distribution in Series MLC MTJ (μA)

FIGURE 3.17
See color insert. Threshold current distributions of resistance state transitions for the series MLC MTJ. (a) Dependent transitions. (b) Independent transitions.

respectively. Similar dependency of the write error rate on the MTJ working region is observed. Interestingly, the minimum write error rate occurs when $T_w = 10$ ns, as the standard deviation/mean ratio reaches its minimum value (see Figure 3.15(b)). Compared to parallel MLC MTJs, series MLC MTJs demonstrate much higher write reliability at the same technology node, while requiring a slightly larger switching current and higher write energy consumption.

3.7 Conclusion

In this work, we analyzed the dependency of the sensing errors and the read disturbance of the STT-RAM cell on the read current under the scaled technology nodes. By taking into account the trade-off between these two major sources of read errors, we also demonstrated the importance of read current optimization for the enhancement of STT-RAM readability. To achieve a reasonable read error rate of STT-RAM at scaled technology nodes, many other circuit and device optimizations must be made instead of taking the current simple scaling path.

We also quantitatively analyzed the impacts of the process variations and the thermal fluctuations on the performance and reliability of both parallel and series multilevel cell (MLC) STT-RAM cell designs. Compared to conventional single-level cell (SLC) STT-RAM designs, the different storage mechanism of the MLC STT-RAM results in very unique operation failure models and reliability optimization concerns. Our results showed that the resistance states of both MLC STT-RAM cells must be optimized in the designs to minimize the read error. The magnitude of the write current must also be carefully selected to suppress both incomplete write and overwrite failures. Our simulation results show that series MLC STT-RAM demonstrates much better reliability in both write and read operations than the parallel MLC STT-RAM under the same fabrication conditions. Also, as expected, the readability is still the biggest concern in both MLC STT-RAM designs.

Acknowledgments

This work is supported in part by National Science Foundation (NSF) grants CCF-1217947, CNS-1311706, and CNS-1116171/1342566.

References

1. http://www.itrs.net/Links/2011ITRS/Home2011.
2. D. Apalkov, S. Watts, A. Driskill-Smith, E. Chen, Z. Diao, and V. Nikitin. Comparison of Scaling of In-Plane and Perpendicular Spin Transfer Switching Technologies by Micromagnetic Simulation. *IEEE Transactions on Magnetics*, 46(6):2240–2243, 2010.
3. L. Berger. Emission of Spin Waves by a Magnetic Multilayer Traversed by a Current. *Physical Review B*, 54:9353–9358, 1996.

4. Y. Cao et al. New Paradigm of Predictive MOSFET and Interconnect Modeling for Early Circuit Design. In *IEEE Custom Integrated Circuit Conference*, 2000, pp. 201–204. http://www-device.eecs.berkeley.edu/ptm.

5. Y. Chen, X. Wang, H. Li, H. Xi, Y. Yan, and W. Zhu. Design Margin Exploration of Spin-Transfer Torque RAM (STT-RAM) in Scaled Technologies. *IEEE Transactions on Very Large Scale Integration (VLSI) Systems*, 18(12):1724–1734, 2010.

6. Y. Chen, X. Wang, W. Zhu, H. Li, Z. Sun, G. Sun, and Y. Xie. Access Scheme of Multi-Level Cell Spin-Transfer Torque Random Access Memory and Its Optimization. In *53rd IEEE International Midwest Symposium on Circuits and Systems*, August 2010, pp. 1109–1112.

7. Y. Chen, W.-F. Wong, H. Li, and C.-K. Koh. Processor Caches Built Using Multi-Level Spin-Transfer Torque RAM Cells. In *International Symposium on Low Power Electronics and Design 2011*, August 2011, pp. 73–78.

8. Z. Diao, Z. Li, S. Wang, Y. Ding, A. Panchula, E. Chen, L. Wang, and Y. Huai. Spin-Transfer Torque Switching in Magnetic Tunnel Junctions and Spin-Transfer Torque Random Access Memory. *Journal of Physics: Condensed Matter*, 19:165209, 2007.

9. G. Fuchs, I.N. Krivorotov, P.M. Braganca, N.C. Emley, A.G.F. Garcia, D.C. Ralph, and R.A. Buhrman. Adjustable Spin Torque in Magnetic Tunnel Junctions with Two Fixed Layers. *Applied Physics Letters*, 86:152509, 2005.

10. T.L. Gilbert. A Lagrangian Formulation of the Gyromagnetic Equation of the Magnetization Field. *Physical Review*, 100(1243), 1955.

11. Y. Huai. Spin-Transfer Torque MRAM (STT-MRAM): Challenges and Prospects. *AAPPS Bulletin*, 18(6):33–40, 2008.

12. S. Ikeda, K. Miura, H. Yamamoto, K. Mizunuma, H.D. Gan, M. Endo, S. Kanai, J. Hayakawa, and H. Matsukura, F. Ohno. A Perpendicular-Anisotropy CoFeB-MgO Magnetic Tunnel Junction. *Nature Materials*, 9(9):721–724, 2010.

13. T. Ishigaki, T. Kawahara, R. Takemura, K. Ono, K. Ito, H. Matsuoka, and H. Ohno. A Multilevel-Cell Spin-Transfer Torque Memory with Series-Stacked Magnetotunnel Junctions. In *Symposium on VLSI Technology*, June 2010, pp. 47–48.

14. C.W. Smullen IV, V. Mohan, A. Nigam, S. Gurumurthi, and M.R. Stan. Relaxing Non-Volatility for Fast and Energy-Efficient STT-RAM Caches. In *11th International Symposium on High-Performance Computer Architecture*, 2011, pp. 50–61.

15. M.-Y. Kim, H. Lee, and C. Kim. PVT Variation Tolerant Current Source with On-Chip Digital Self-Calibration. *IEEE Transactions on Very Large Scale Integration Systems*, 20(4):737–741, 2012.

16. J. Li, H. Liu, S. Salahuddin, and K. Roy. Variation-Tolerant Spin-Torque Transfer (STT) MRAM Array for Yield Enhancement. In *IEEE Custom Integrated Circuits Conference*, September 2008, pp. 193–196.

17. X. Lou, Z. Gao, D.V. Dimitrov, and M.X. Tang. Demonstration of Multilevel Cell Spin Transfer Switching in MgO Magnetic Tunnel Junctions. *Applied Physics Letters*, 93(24):242502, 2008.

18. N. Mojumder, C. Augustine, D. Nikonov, and K. Roy. Effect of Quantum Confinement on Spin Transport and Magnetization Dynamics in Dual Barrier Spin Transfer Torque Magnetic Tunnel Junctions. *Journal of Applied Physics*, 108(10):104306–104312, 2010.

19. K. Munira, W.A. Soffa, and A.W. Ghosh. Comparative Material Issues for Fast Reliable Switching in STT-RAMs. In *11th IEEE Conference on Nanotechnology*, August 2011, pp. 1403–1408.

20. A. Nigam, C.W. Smullen IV, V. Mohan, E. Chen, S. Gurumurthi, and M.R. Stan. Delivering on the Promise of Universal Memory for Spin-Transfer Torque RAM (STT-RAM). In *Proceedings of the 17th IEEE/ACM International Symposium on Low-Power Electronics and Design (ISLPED '11)*, Piscataway, NJ, 2011, pp. 121–126.

21. A. Raychowdhury, D. Somasekhar, T. Karnik, and V. De. Design Space and Scalability Exploration of 1T-1STT MTJ Memory Arrays in the Presence of Variability and Disturbances. In *IEEE International Conference on Electron Devices Meeting*, December 2009, pp. 1–4.

22. R. Sbiaa, R. Law, S.Y.H. Lua, E.L. Tan, T. Tahmasebi, C.C. Wang, and S.N. Piramanayagam. Spin Transfer Torque Switching for Multi-bit per Cell Magnetic Memory with Perpendicular Anisotropy. *Applied Physics Letters*, 99(9):092506, 2011.

23. J.Z. Sun. Spin-Current Interaction with a Monodomain Magnetic Body: A Model Study. *Phys. Rev. B*, 62:570–578, 2000.

24. Z. Sun, H. Li, Y. Chen, and X. Wang. Variation Tolerant Sensing Scheme of Spin-Transfer Torque Memory for Yield Improvement. In *IEEE/ACM International Conference on Computer-Aided Design*, November 2010, pp. 432–437.

25. Y. Zhang, Y. Li, A.K. Jones, X. Wang, and Y. Chen. Asymmetry of MTJ Switching and Its Implication to the STT-RAM Designs. Presented at Design Automation and Test in Europe, March 2012.

26. W. Zhao and Y. Cao. New Generation of Predictive Technology Model for Sub-45 nm Early Design Exploration. *IEEE Transactions on Electron Devices*, 53(11):2816–2823, 2006.

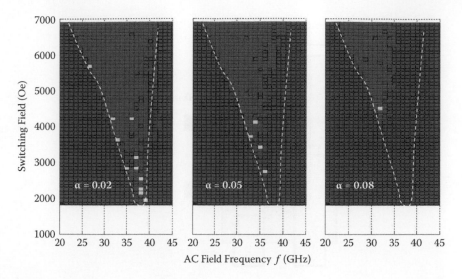

FIGURE 1.13
Calculated switching maps for a segmented grain with two exchange break layers inserted. The reversal field is a 1 ns pulse at the direction 30° off the grain anisotropy easy axis, while a steady circular ac field is applied in the horizontal plane. A red cell indicates irreversible magnetization switching, and blue otherwise. The switching field of the grain without the ac field is 12,000 Oe.

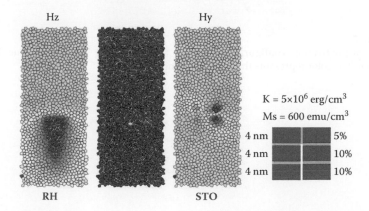

FIGURE 1.14
Map of the perpendicular (left) and down-track (right) components of the recording field along with the resulting transition pattern (middle). The head moves down relative to the medium. x is in the cross-track direction, y the down-track direction, and z the perpendicular direction.

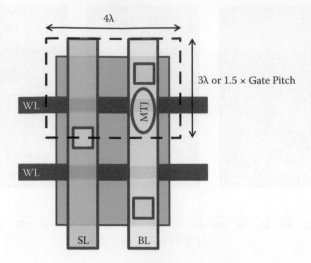

FIGURE 2.10
A representative 1T-1MTJ bitcell for the array architecture in Figure 2.9. λ is assumed to be the minimum metal width. The bitcell size is limited by the minimum metal pitch or 1.5 times the gate pitch.

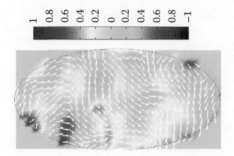

FIGURE 2.27
Transient magnetization configurations during STT switching for an in-plane MTJ of 160×80 nm. The color represents the M_z component.

FIGURE 2.29
Formation of a magnetization vortex at the end of the STT switching process.

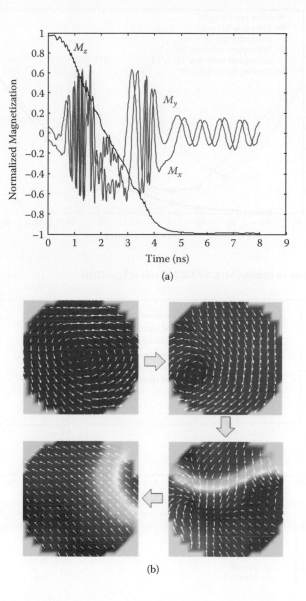

(a)

(b)

FIGURE 2.30
(a) Simulated volume-averaged magnetization components as a function of time for a pMTJ (80 nm in diameter). (b) The corresponding transient magnetization configurations during STT switching. The color represents the M_z component.

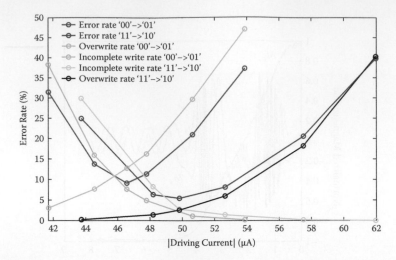

FIGURE 3.13
Writing error rate in parallel MLC STT-RAM cell at $T_w = 10$ ns.

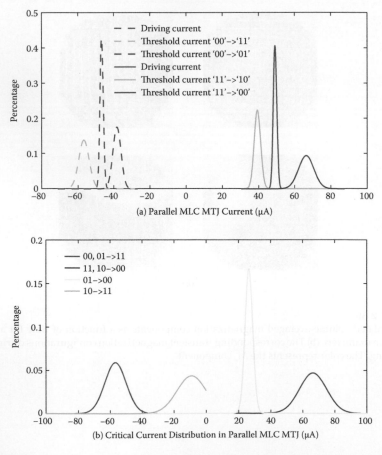

FIGURE 3.15
Threshold current distributions of resistance state trasitions for the parallel MLC MTJ. (a) Dependent transitions. (b) Independent transitions.

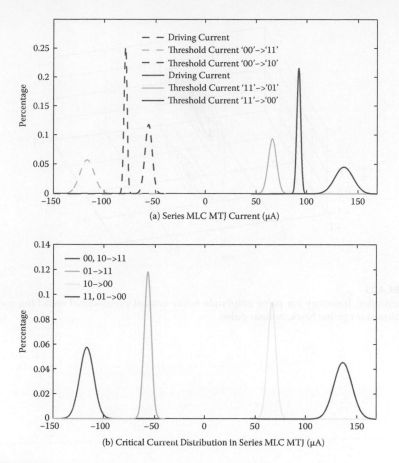

FIGURE 3.17
Threshold current distributions of resistance state transitions for the series MLC MTJ. (a) Dependent transitions. (b) Independent transitions.

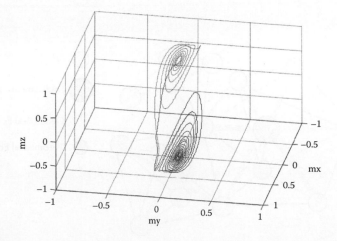

FIGURE 4.22
Magnetization trajectory for pulse amplitude reaching critical precessional switching current. Red, ultra-short pulse; black, normal pulse.

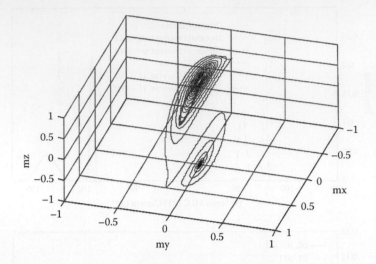

FIGURE 4.23
Magnetization trajectory for pulse amplitude below critical precessional switching current. Red, ultra-short pulse; black, normal pulse.

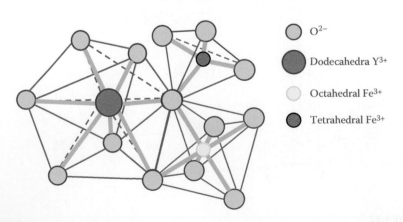

O^{2-}

Dodecahedra Y^{3+}

Octahedral Fe^{3+}

Tetrahedral Fe^{3+}

FIGURE 5.1
Schematic diagram of three polyhedron sites in YIG crystal.

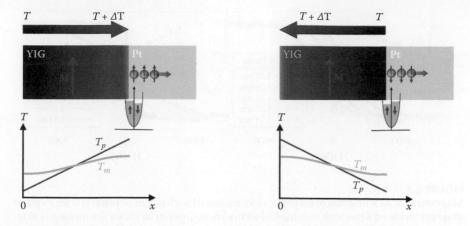

FIGURE 5.23
(a) The phonon temperature is higher than the magnon temperature at the interface, which enables the spin injection with up-polarization from the magnetic layer to the NM layer. This process causes a pumping effect on magnetization precession in YIG. (b) The phonon temperature is lower than the magnon temperature at the interface, which enables the spin injection with down-polarization from the magnetic layer to the NM layer. This process causes a pumping effect on the magnetization precession in YIG.

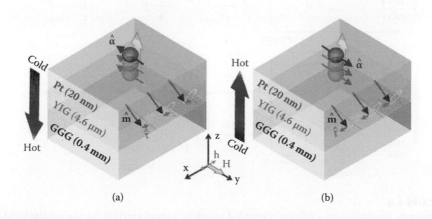

FIGURE 5.24
Experimental configuration. Graphs (a) and (b) show the situations for two difference temperature gradients. (Reprinted with permission from L. Lu, et al., *Phys. Rev. Lett.* 108, 257202 (2012). Copyright 2012 by the American Physical Society.)

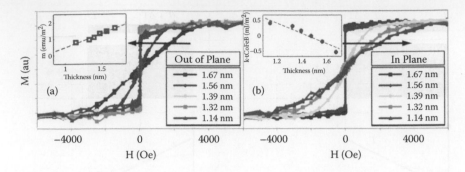

FIGURE 6.5
Magnetization as a function of magnetic field, measured with (a) out-of-plane and (b) in-plane magnetic fields on films with varying CoFeB thickness. Inset in (a) shows the measured magnetic moment as a function of film thickness. Inset in (b) shows the dependence of perpendicular anisotropy on thickness, consistent with an interfacial origin of the perpendicular anisotropy. (From P. K. Amiri, et al., *Applied Physics Letters*, 98, 2011. With permission.)

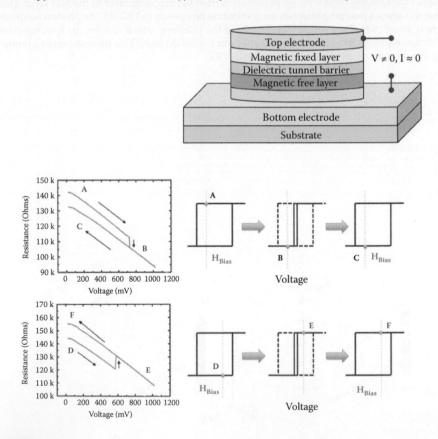

FIGURE 6.8
Schematic illustration of a voltage-controlled magnetic tunnel junction (top) using the VCMA effect [87, 107, 108]. Bottom panel shows the voltage-induced switching of a nanoscale device with in-plane magnetization. The switching process is assisted by a small magnetic field that also determines the switching direction. Note that the same voltage polarity is used for switching in both directions.

4

Spintronic Device Memristive Effects and Magnetization Switching Optimizing

Xiaobin Wang

Avalanche Technology, Fremont, California

CONTENTS

4.1 Introduction

Spintronics, an exploration of spin properties instead of or in addition to charge degrees of freedom, holds promise to continue device miniaturization in the postsilicon era and beyond the age of Moore's law. The benefits of spintronics come from the fact that there is no need for electric current to retain spin, and there are many paths to change spin without massively moving electrons.

While full of fundamental challenges, spintronics technology has been tested in existing applications, such as magnetic data storage devices and field-driven magnetic random access memory (MRAM). Emerging spintronics applications, such as spin torque random access memory (SPRAM), spin-based logic, spin transistors, and spin quantum computer, are currently under extensive research and development. For example, SPRAM is an emerging technology that integrates magnetic tunneling junction (MTJ) and complementary metal oxide semiconductor (CMOS) to achieve nonvolatility, fast writing/reading speed, almost unlimited programming endurance, and zero standby power. As charge-based memory technologies such as dynamic random access memory (DRAM) or NAND Flash are facing severe scalability problems due to precise charge placement and sensing hurdles in deep-submicron processes, SPRAM is widely believed in industry to be a candidate to replace charge-based memory. A recent announcement of the first commercial SPRAM (Everspin, November 2012) and continuing demonstration of SPRAM scalability (down to the 54 nm technology node) have confirmed the viability of the SPRAM technology path.

Electric field- or voltage-induced magnetization switching provides an exciting opportunity for switching magnetization without resorting to electronic transport across the device. It has been demonstrated both experimentally and theoretically that electric fields can manipulate magnetic anisotropy to induce magnetization switching. Another electric field-induced magnetization switching system is multiferroics, which has coupled magnetic and ferroelectric order parameters. When applied to practical devices, electric field-induced magnetization switching can lead to low-power magnetization switching with very little heat generation.

Another intensively investigated approach to switch magnetization is with ultra-fast light pulse. It is generally believed that ultra-fast laser pulse could either rapidly heat the spin system to Curie temperature for easy magnetization switching or induce an inverse Faraday effect to switch magnetization effectively. One of the advantages of laser pulse-induced magnetization switching is the extremely fast magnetization switching speed. Magnetization speeds down to the picosecond level have been reported utilizing this switching mechanism.

The modern nanoscale spintronic device has reached a stage where device performance is explicitly determined by coupling between macroscopic magnetization dynamics, microscopic spin electron transport, and finite temperature thermal agitations. An ever-increasing challenge is to achieve well-controlled faster magnetization switching and maintain stable magnetization states at longer timescales at higher temperatures. A coupled, multiscale dynamic approach considering both device mean performance and its variance is required for spintronic devices' understanding and design.

Memristive effects are universal for spin torque spintronic devices at the timescale that explicitly involves the interactions between magnetization dynamics and electronic charge transport. In this chapter, we explore spintronic device memristive effects through stochastic magnetization dynamics. Although memristive effects provide many exciting new application possibilities for spintronic devices, they also have quite strict constraints on spintronic device operations, such as writing speed and heat dissipation. We will explore spintronic device write and read optimization in light of spintronic memristive effects.

4.2 Spintronic Device Memristive Effects

4.2.1 Spin Torque Device Read and Write Dynamics

The giant magnetoresistive (GMR) effect was the key physical principle for commercial magnetic hard disk drive read sensors, for which the Nobel Prize in Physics was awarded to Peter Grunberg and Albert Fert. In GMR devices, two ferromagnetic layers sandwich a nonferromagnetic metal known as a spin valve structure. Spin valve resistance depends on the relative magnetic orientations of the two ferromagnetic layers. Device resistance is low when the magnetizations of two ferromagnetic layers are parallel, and device resistance is high when the magnetizations of two ferromagnetic layers are antiparallel. A GMR-based spin valve has a dramatically increased hard disk drive areal density through its excellent sensitivity for magnetic flux. Recently the GRM structure has been replaced by the tunneling magnetoresistance (TMR) structure. The TMR device has two ferromagnetic layers that sandwich a tunneling-insulating barrier known as the magnetic tunneling junction (MTJ). Similar to the spin valve, MTJ resistance is low when the magnetizations of two ferromagnetic layers are parallel, and device resistance is high when the magnetizations of two ferromagnetic layers are antiparallel. Compared to GMR, TMR has a much higher value of resistance difference over resistance. The TMR value, which is usually defined as the difference between high resistance and low resistance divided by low resistance, is around 100% in commercial sensors and could reach 600 to 1000% in the lab.

The key physics principle behind GMR and TMR sensing operations is the resistance dependence upon magnetization states. It can be written as

$$G(\theta) = \frac{G_P + G_{AP}}{2} - \frac{G_{AP} - G_P}{2}\cos\theta \qquad (4.1)$$

where G_P, G_{AP} are the conductances for free layer magnetization parallel and antiparallel to reference layer magnetization. θ is the angle between free layer magnetization and reference layer magnetization.

For magnetic random access memory, besides providing a sensing scheme based upon magnetization state-dependent resistance, a spin valve or MTJ

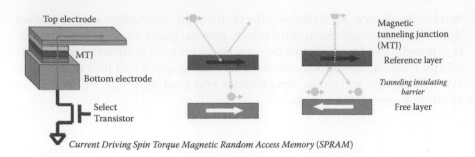

FIGURE 4.1

Schematic pictures of structure and working principle of spin torque random access memory. (From X. Wang et al., *CMOS Processors and Memories*, Springer, Berlin, 2010. With permission.)

structure also stores magnetic information. Magnetic tunneling junction (MTJ) is the key component of spin torque random access memory (SPRAM) cell. Figure 4.1 shows a schematic picture of MTJ-based SPRAM. Two ferromagnetic layers sandwich an oxide barrier layer, for example, MgO. The magnetization direction of one ferromagnetic layer (reference layer) is fixed by coupling to a pinned magnetization layer above, while the magnetization direction of the other ferromagnetic layer (free layer) can be changed. The 0/1 data information is stored as the magnetization direction of the free layer in SPRAM.

The writing operation of SPRAM is based upon spin polarization current-induced magnetization switching [1, 2]. To switch the free layer magnetization to the same direction of the reference layer magnetization, an electric current passes from the reference layer to the free layer. The injected current electrons have spins pointing to the same and opposite directions of the reference layer magnetization. After passing through the reference layer, the electrons have a preferred spin orientation direction pointing to the same direction of the reference layer magnetization. This is because most of the electrons with spin pointing to the opposite direction of the reference layer magnetization are reflected back due to interaction between the itinerant electron spin and the reference layer local magnetization. The polarized current electrons, with a net spin moment in the same direction as the reference layer magnetization, will switch the free layer magnetization to the same direction as the reference layer magnetization.

In order to switch the free layer magnetization to the opposite direction of the reference layer magnetization, electron current passes from the free layer to the reference layer. Based upon the same physics argument as before, the reflected electrons from the reference layer have preferred spin direction opposite to the direction of the reference layer magnetization. These will switch the free layer magnetization to the opposite direction of the reference layer magnetization.

The key physical principle behind MTJ writing is that the free layer macroscopic magnetization states are changed by the torque provided through spin current. This is a complex dynamic process described by the stochastic

Landau-Lifshitz-Gilbert equation at a finite temperature with spin torque terms:

$$\frac{d\bar{m}}{dt} = -\bar{m} \times (\bar{h}_{eff} + \bar{h}_{th}) - \alpha\bar{m} \times \bar{m} \times (\bar{h}_{eff} + \bar{h}_{th}) + \frac{\bar{T}}{M_s}$$

(4.2)

where $\bar{m}$ is the normalized magnetization vector, and time t is normalized by γM_s, with γ being the gyromagnetic ratio and M_s being the magnetization saturation.

$$\bar{h}_{eff} = \bar{H}_{eff} / M_s = \frac{\partial \varepsilon}{\partial \bar{m}}$$

is the normalized effective magnetic field with normalized energy density ε, and α is the damping parameter. $\bar{h}_{fluc}$ is the thermal fluctuation field, whose magnitude is determined by the fluctuation-dissipation condition at room temperature and whose formalism follows reference [3].

$$\bar{T}_{norm} = \frac{\bar{T}}{M_s V}$$

is the normalized spin torque term with units of magnetic field. The net spin torque $\bar{T}$ can be obtained through the microscopic quantum electronic spin transport model [4–11]. At the macroscopic magnetization dynamics level, spin torque can be approximated through an adiabatic term proportional to $\bar{m} \times \bar{m} \times \bar{p}$ and a nonadiabatic term proportional to $\bar{m} \times \bar{p}$, where $\bar{p}$ is a unit vector pointing to the spin polarization direction.

Let's consider a MTJ free layer element with a perpendicular uniaxial anisotropy energy $\varepsilon = (\sin^2 \theta) / 2$, where θ is the angle between magnetization orientation and anisotropy axis. The magnetic energy is normalized by $M_s H_k V$, with H_k being the magnetic anisotropy, M_s being the magnetic saturation, and V being the element volume. Magnetization dynamics based upon the stochastic Landau-Lifshitz-Gilbert equation with the Berger and Slonczewski spin torque term (4.2) in this case can be written as [12]

$$d\theta = -\frac{\partial \varepsilon}{\partial \theta} dt + h\sin\theta dt + \delta ctg\theta dt + \sqrt{2\delta} dB$$

(4.3)

$$= -\sin\theta\cos\theta dt + h\sin\theta dt + \delta ctg\theta dt + \sqrt{2\delta} dB$$

where time is normalized by $\alpha\gamma H_k$, with γ being the gyromagnetic ratio and α the damping parameter. h is the effective field whose magnitude is proportional to the spin current magnitude.

$$\delta = \frac{k_B T}{H_k M_s V}$$

is the thermal energy over magnetic energy. $k_B T$ is the Boltzmann thermal factor. dB in Equation (4.3) represents Brownian motion increment.

The current magnitude required to switch free layer magnetization is a dynamic concept. Although we talk about the critical switching current h_{cr}, which is the minimum current magnitude achieving $\inf_h \left\{ \sin \theta (h - \cos \theta) \geq 0 \right\}$ for all magnetization angles θ, magnetization switching could happen at any current magnitude due to the combined spin torque dynamics and random thermal fluctuations. An example of current magnitude versus switching time is shown in Figure 4.2. This curve shows the spin torque-induced magnetization switching in the whole time range, from short-time dynamical switching to long-time thermal switching. Besides fast dynamical magnetization switching induced by a spin current with a magnitude bigger than that of the critical switching current, magnetization switching also occurs in the thermal switching region where spin current magnitude is lower than critical switching current. This curve shows one of the fundamental challenges of spin torque devices, that is, how to achieve fast nanosecond dynamic magnetization switching and at the same time maintain long-time thermal stability at second-to-year scales.

The critical switching current for Equation (4.3) is $h_{cr} = 1$ according to $\inf_h \left\{ \sin \theta (h - \cos \theta) \geq 0 \right\}$. For dynamical switching with current amplitude

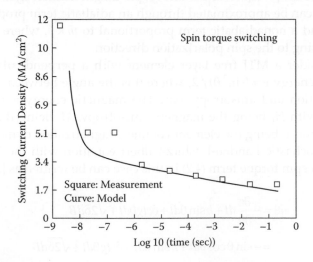

FIGURE 4.2
Dynamic thermal switching of magnetic elements with spin torque current. Squares are experiment data and curves are theoretical calculations. (From X. Wang et al. *IEEE Trans. Magn.*, 45, 2039, 2009. With permission.)

bigger than that for the critical switching current, $h > 1$, the following asymptotic solution of Equation (4.3) can be obtained for current pulse duration versus current amplitude based upon $\delta \ll 1$. $\delta \ll 1$ holds true for a stable magnetic device with magnetic energy bigger than thermal energy at finite temperature:

$$T(\theta,h,\delta) = \frac{(h+1)\ln[\tan(\theta/2)] - \ln[(h+1)\tan(\theta/2)^2 + (h-1)]}{(h-1)(h+1)}$$

$$- \frac{\ln[\theta\sqrt{(h-1)/(2\delta)}] + Ei[\theta^2(h-1)/(2\delta)]/2 + \gamma/2}{h-1} + \frac{\ln(\theta)}{h-1} \qquad (4.4)$$

$$- \frac{\ln(2h)}{(h-1)(h+1)} + \frac{\ln[\sqrt{(h-1)/(2\delta)}]}{h-1} + \frac{\gamma}{2(h-1)}$$

Equation (4.4) is the average time duration T required for a constant amplitude current h to drive magnetization starting at angle θ to the magnetization switching at saddle position.

$$\theta = \left(\int \theta \sin\theta e^{-\sin^2\theta/2\delta} \, d\theta \right) / \left(\int \sin\theta e^{-\sin^2\theta/2\delta} \, d\theta \right)$$

Figure 4.3 shows current pulse magnitude versus inverse of current pulse duration for devices with different thermal stabilities. The magnetization initial angle is determined by the Boltzmann distribution at equilibrium: θ. Linear dependence of current amplitude upon the inverse of current

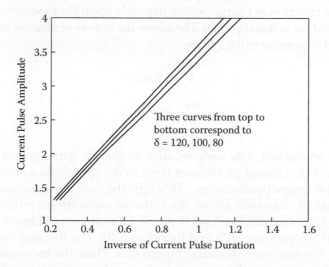

Three curves from top to bottom correspond to $\delta = 120, 100, 80$

FIGURE 4.3
Spin torque current magnitude versus inverse of the pulse duration to achieve MTJ magnetization switching.

duration is observed. Figure 4.3 shows clearly that in order to achieve fast magnetization switching, a sufficient amount of the charge (which is proportional current amplitude multiplied by current duration) is required to flow through the MTJ device. It is the time integration of the current, instead of current amplitude, that determines the MTJ magnetization final state. Because MTJ resistance depends upon the magnetization state, as in Equation (4.1), the MTJ final resistance state depends upon the charge passing through the device.

4.2.2 Spintronic Device Memristive Effects

A device with resistance dependence upon charge or current integration was proposed as a memristor in [13]. The original definition of memristor was

$$d\phi = Mdq \tag{4.5}$$

where q is the electric charge, ϕ is the magnetic flux, and M is the memristance. Because magnetic flux is the integration of voltage and charge is the integration of current, it is easy to show that

$$M(q) = \frac{d\phi/dt}{dq/dt} = \frac{V}{I} \tag{4.6}$$

What makes memristance different from an ordinary constant resistance or even a current- or voltage-dependent nonlinear resistance is that memristance is a function of charge, which depends upon the hysteretic behavior of the current (or voltage) profile. The above definition of memresistor can be generalized to a memristive system:

$$V = M(m)i$$
$$\frac{dm}{dt} = f(i,m) \tag{4.7}$$

where m is an implicit state variable, such as charge, that depends upon the integration of the current profile over time. In the spintronic memristor here, m can be the magnetization state. Although the unit of memristance is the same as that of resistance (ohm), the intrinsic constitution relation for the memristor is quite different from three other well-known circuit elements: resistance (R), capacitance (C), and inductance (L). The intrinsic constitution relation for the memristor is charge versus flux. Thus, the memristor has the unique ability to accumulate current/voltage information through constant current or voltage driving strength. As a comparison, in order to keep accumulating charge, the capacitor must be driven with a varying voltage. This

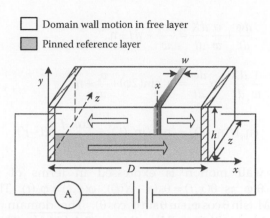

FIGURE 4.4
Spin valve with domain.

is because the intrinsic constitution relation of the capacitor is charge versus voltage. Because of its unique intrinsic constitution, the memristor is also called the fourth element of the circuit.

It has been shown in [14] that memristive effects are universal for spin torque spintronic devices at the timescale that explicitly involves the interactions between magnetization dynamics and electronic charge transport. For spin torque devices, the device resistance depends upon its magnetization state. The magnetization state of the device depends upon the cumulative effects of electron spin excitations. Thus, a spin torque spintronic device is a spintronics memristor with its resistance dependent upon the integral effects of its current profile.

Another example of a spintronic memristor is the spintronic domain wall device. The device structure is shown in Figure 4.4. It consists of a long spin valve strip that includes two ferromagnetic layers: the reference layer and free layer. The magnetization direction of the reference layer is fixed by coupling to a pinned magnetic reference layer. The free layer is divided by a domain wall into two segments that have opposite magnetization directions to each other. The device domain resistance depends upon the domain wall position as

$$R(X) = \left[\frac{R_H}{D} - \frac{(R_H - R_L)X(t)}{D} \right] \qquad (4.8)$$

where R_H and R_L are the high and low resistance of the spin valve, D is the spin valve length, and $X(t)$ is the domain wall position. The free layer magnetization dynamics can be described by the stochastic Landau-Lifshitz-Gilbert equation using rigid wall approximation [15, 16]:

$$\frac{d\varphi}{dt} + \frac{\alpha}{w}\frac{dX}{dt} = \frac{\beta v_s}{w} + \gamma H + \eta_\varphi$$

$$\frac{1}{w}\frac{dX}{dt} - \alpha\frac{d\varphi}{dt} = \omega_0 \sin(2\varphi) + \frac{v_s}{w} + \eta_X \qquad (4.9)$$

$$\left\langle \eta_\varphi(t)\eta_\varphi(t')\right\rangle = \left\langle \eta_X(t)\eta_X(t')\right\rangle = \frac{2\alpha k_B T}{\hbar N}\delta(t-t')$$

The domain wall motion is expressed in terms of magnetization spherical angle θ, φ as $\theta(x,t) = \theta_0(x - X(t))$, $\varphi(x,t) = \varphi_0(t)$. The magnetization vector is $M_s(\sin\theta\cos\varphi, \sin\theta\sin\varphi, \cos\theta)$. The domain wall shape is $\cos\theta_0(x) = \tanh(x/w)$ with a wall length $w = \sqrt{2A/M_s H_k}$. The wall length is determined by M_s exchange strength A, and easy-axis anisotropy H_k in the x-direction. $\omega_0 = \gamma H_p/2$, with γ being the gyromagnetic ratio and H_p being a hard y plane anisotropy. $v_s = PJu_B/eM_s$ is the spin torque excitation strength. Spin torque excitation strength is proportional to current density J. u_B is the Bohr magneton and e is the elementary electron charge. P is the polarization efficiency, which can be approximate as

$$P = \frac{\sigma_\uparrow - \sigma_\downarrow}{\sigma_\uparrow + \sigma_\downarrow}$$

for two spin channel models with spin-up and -down conductivity $\sigma_{\uparrow,\downarrow}$. H is the effective magnetic field. β is the ratio of nonadiabatic spin torque strength to adiabatic spin torque strength.

Figure 4.5 shows the inverse of current pulse duration versus current pulse amplitude for a domain wall to travel a fixed distance. The inverse of the current pulse is normalized by v_s/w. The current pulse amplitude is normalized by $PJu_B/eM_s w\omega_0$, and the temperature is normalized by $4k_B T/N\hbar\omega_0$. The fixed distance is 1 (normalized by w). The inverse of the current pulse duration depends almost linearly upon the current pulse amplitude for a wide range of temperatures. For the spin torque domain wall device, it is the charge instead of current magnitude that determines the final domain wall position and the device resistance. For the domain wall device, a continuum of resistance states can be explored through moving the domain wall across the thin-film structure.

4.3 Potential Application of Spintronic Memristive Effects

After proposing resistive memory switching in the metal-insulator-metal structure as nanoscale memristors [17], there is a burst of interest in the potential

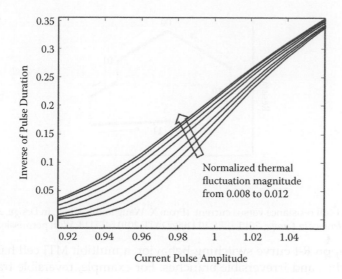

FIGURE 4.5
Inverse of current pulse duration versus pulse amplitude for a domain wall to move a given distance at different temperatures. (From X. Wang et al., *IEEE Device Lett.* 31, 20, 2010. With permission.)

application of nanoscale memristors. It is also shown that the majority of memory material-based two-terminal electronic devices behave as memristors or a combination of memristors, memcapacitors, and meminductors when subjected to time-dependent perturbations [18]. Thus, it is no surprise that there are many potential applications of spintronic memristive effects.

Random access memory is the most obvious application of memristive systems. For spin torque random access memory, one may ask why it is necessary to examine spin torque-induced magnetization switching from a memristive point of view. The benefit of investigating memristive effects on spin torque random access memory can be illustrated through switching of a multicell SPRAM. It was shown [19] that memristive effects of magnetization switching provide additional freedom for us to manipulate magnetization switching between multiple equilibrium states.

MTJ multilevel cell spin torque switching has been demonstrated experimentally in [20]. Two multibit MTJ cell examples are presented. In the first example, two horizontal domains are introduced in the MTJ free layer to generate four bits: 00/01/10/11. In the second example, two MTJs are stacked vertically to generate four bits. Figure 4.6 shows schematically the switching behavior of these multibit MTJ cells through the R-I curve. One bit of the two-bit cell is hard and requires higher current to switch, and the other bit is soft and requires less current to switch. Thus, by varying switching current magnitude, individual bits can be switched. In the following we denote the first bit as the hard bit and the second bit as the soft bit.

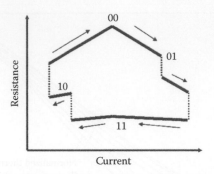

FIGURE 4.6
Multibit MTJ cell resistance versus current. (From X. Wang and Y. Chen, in *Design, Automation and Test Conference in Europe*, 2010, IEEE 10.1109/DATE.2010.5457118. With permission.)

Based upon R-I curve switching behavior, a multibit MTJ cell has reversible branches and irreversible branches. For example, reversible transitions between 11 and 10, 01 and 00 can be achieved; however, the 00 state cannot be directly switched to the 10 state using a single current pulse because switching the first bit requires higher current than switching the second bit.

MTJ memristive effects provide an approach to design a two-bit MTJ cell with reversible switching for all bit combinations. As illustrated in the previous section, spin torque-excited MTJ is a memristor. When the observation time reaches a scale that explicitly involves interactions between magnetization dynamics and electronics transport, the I-V curve (or widely used R-I curve) is not intrinsic to MTJ. MTJ resistance switching is determined by the current/voltage profile. MTJ switching behavior is determined by integration of the current pulse instead of the current magnitude itself.

Two curves in Figure 4.7 correspond to two MTJ subcells with the same material. The two cells have different shape-induced anisotropy and effective volume. The dot line subcell has a more square shape and bigger surface area, while the square curve has a more elongated shape and less surface area. The elongated subcell requires less current to switch at a smaller current magnitude; however, the switching speed of the elongated subcell is slower than that of the squared cell when the current magnitude increases. As a result, two switching curves cross at a current magnitude of around 900 µA and pulse width of around 4 ns. If we denote the elongated subcell as a soft bit, the two subcells of MTJ can be switched freely by using currents with different pulse widths. The longer current pulse width switches the soft subcell, corresponding to switching between 00 and 01 states. The shorter current pulse width switches the hard subcell, corresponding to switching between 00 and 10 states.

It is well known that a memristor has rich dynamic behavior when excited with a dynamic current/voltage profile. The example here shows a practical application in multibit MTJ cell switching. By controlling the switching

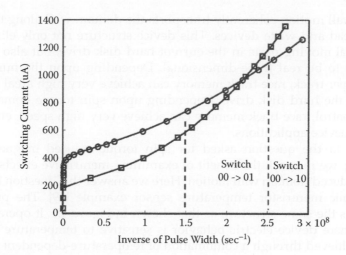

FIGURE 4.7
Switching current versus inverse pulse width for two multibit MTJ subcells' pulse magnitude versus duration for moving domain wall to a given distance. (From X. Wang and Y. Chen, in *Design, Automation and Test Conference in Europe*, 2010, IEEE 10.1109/DATE.2010.5457118. With permission.)

current pulse width, the subcells of MTJ are designed to switch freely between different bit combinations. What is illustrated here is that memristive effects of magnetization switching provide additional freedom to manipulate magnetization switching.

Crossbar architecture was investigated intensively on resistive memory structure [21]. Compared to conventional memory structure, cross-point design could relax the density limitation of two terminal devices imposed by the CMOS circuit, thus offering a significant improvement in achievable density. Similar architecture was proposed for spin torque devices and MRAM structures [22]. Other areas of memristor digital applications include logic [23], field-programmable gate arrays (FPGAs) [24], image processing [25], etc. It would be interesting to see future spintronic devices penetrate into these technology applications.

A nonvolatile full adder based on logic-in-memory architecture was fabricated using magnetic tunnel junctions (MTJs) in combination with metal oxide semiconductor (MOS) transistors [26]. MTJs are used for TMR read and spin torque magnetization switching write. The MOS transistors are fabricated using a complementary metal oxide semiconductor (CMOS) process. The basic operation of the full adder was demonstrated. An all-function-in-one magnetic chip reconfigurable logic device was proposed based upon an MTJ stack with spin torque magnetization [27].

Race track memory [28] consisting of three-dimensional ferromagnetic nanowires that are encoded with domain walls has been proposed to achieve large storage capability and fast write read speed. Spin torque-induced domain wall motion coherently transports the domain wall along the wire to pass read and write devices. This device structure not only eliminates mechanical moving parts in the current hard disk drive, but also has the capability to be real three-dimensional. Depending upon the number of domains per track, race track memory can achieve very high areal density, replacing the hard disk drive. Depending upon spin torque domain wall motion control, race track memory can achieve very high speed, encroaching logic device applications.

Similar to the question asked for spin torque-induced magnetization switching, we may ask the benefit of examining memristive effects in spin torque-induced domain wall motion. Here we answer this question through a spintronic memristor temperature sensor example [29]. The proposed device has the same structure as that shown in Figure 4.4. It operates in a region where device electric behavior is sensitive to temperature change. This is achieved through a combination of temperature-dependent domain wall mobility and positive feedback between resistance and driving strength in the memristor.

A biasing voltage pulse with constant magnitude is applied to the spintronic memristor. Resistances before and after voltage pulse are measured. This resistance difference is calibrated to sense temperature magnitude. Solid curves in Figure 4.8 show the resistance as a function of current pulse duration at different temperatures. Positive feedback between the resistance and domain wall driving strengths is critical in the observed high sensitivity in Figure 4.8.

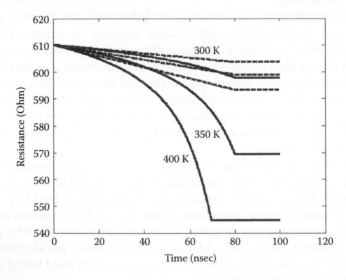

FIGURE 4.8

Spintronic memristor resistance as a function of time at different temperatures for a constant magnitude voltage pulse driving. (From X. Wang et al., *IEEE Device Lett.* 31, 20, 2010. With permission.)

The positive feedback between resistance and driving strength is a unique property of the memrisor. Memristor's resistance depends upon the integration of current/voltage excitation. For a constant voltage pulse driving, higher temperature results in an increased domain wall moving distance. The increased domain wall moving distance results in a smaller resistance. The smaller resistance results in a higher driving current density, thus providing positive feedback to further increase domain wall distance. This positive feedback accelerates domain wall speed and reduces device resistance further for constant voltage pulse driving.

For a comparison, dash curves in Figure 4.8 are resistance change for non-memristive devices without positive feedback between resistance change and driving strength. The dash curve is generated with a fixed domain wall driving strength.

For the device design example in Figure 4.8, the material properties are magnetization saturation $M_s = 1010$ emu/cc, exchange strength $A = 1.8 \cdot 10^{-11}$ J/m, easy z-axis anisotropy $H_k = 100$ Oe, and hard y-axis anisotropy $H_p = 5000$ Oe. The resistance of square thin film is 50 Ω for 70 Å thickness square thin film, and GMR is 12%. The geometry of the device is 268 nm long, 10 nm wide, and 1.7 nm thick. The damping parameter is $\alpha = 0.02$ and polarization efficiency is $P = 0.3$. The number of spins in the domain wall is $N = 2.2 \cdot 10^7$. Based upon the above parameters, the normalized thermal magnitude is $4k_BT / N\hbar\omega_0 = 0.008$ at room temperature $T = 300$ K. The normalized current driving strength is $PJu_B / eM_sw\omega_0 = 0.9$ at $J = 2.75 \cdot 10^8$ A/cm^2. A constant 0.3 V voltage pulse with 80 ns duration is applied to the device.

It should be pointed out that the proposed device requires much lower power supply voltage and power consumption compared to the available on-chip temperature sensor. The device could target the highly integrated on-chip thermal detection application, different from many of the stand-alone or bulk thermal sensors.

The feedback between resistance and integrated current or voltage in the memristor provides opportunities for power monitoring and control. Connecting a memristor with a very small resistance could monitor circuit power consumption. This is because the memristor has the ability to accumulate current/voltage through constant current or voltage driving strength. For active circuit power control, instead of a positive feedback between decreasing resistance and increasing domain wall driving strength for a domain wall moving in a direction to reduce resistance, we can design a spintronic device with a negative feedback between resistance and domain wall driving strength for a domain wall moving in a direction to increase resistance. When the circuit power increases, the memristor domain wall is pushed toward a direction to increase the memristor resistance. For memristor resistance comparable to that of the circuit, this reduces the current and power through the circuit. Again, all these spintronic memristor applications could target highly integrated on-chip due to the mature integration between magnetic MTJ and CMOS circuits.

Analog and analog-digital mixture are important areas for memristor application. Neural computing and various learning algorithms are typical examples. Self-organized computation with unreliable memristive nanodevices were shown in [30]. Analog applications explore the continuous states of the memristive devices. Massively parallel computing architecture based upon a memristive system was shown to be able to solve a maze problem faster than current existing algorithms [31]. Other examples include fuzzy logic [32, 33], adaptive filter [34], chaotic circuits [35], etc.

Utilizing memristors as artificial synapses to efficiently realize some basic functions of the brain is an interesting topic. A design scheme is proposed for ultra-low-power neuromorphic hardware with a spin-based device for neurons [36]. Spin neurons are lateral spin valves and domain wall magnets that operate at ultra-low voltage. MTJs are employed for interfacing the spin neurons with charge-based devices, like CMOS, for large-scale networks. The "brain in a box" concept was proposed based upon spintronic devices [37]. The stochastic nature of spin torque switching of MTJs was utilized to implement spike timing-dependent plasticity of synapses. An experimental and theoretical study of MTJ magnetization switching mean and variance was given in [38], and their implications to SPRAM scaling down were presented in [39]. Figures 4.9 and 4.10 show measured and modeled MTJ switching mean and variations at different pulse durations.

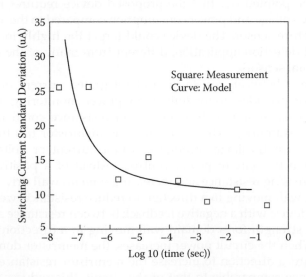

FIGURE 4.9

Switching current density standard deviation versus mean switching time. The mean switching current density is shown in Figure 4.2. (From X. Wang et al., *IEEE Trans. Magn.* 45, 2038, 2009. With permission.)

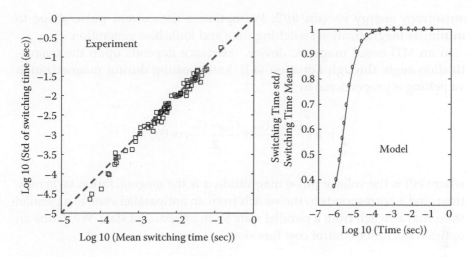

FIGURE 4.10
Modeled and measured switching time standard deviation as a function of the mean switching time. (From X. Wang et al., *IEEE Trans. Magn.* 45, 2038, 2009. With permission.)

4.4 Spintronic Device Writing Optimizing

Although spintronic memristive effects provide many exciting new application possibilities, they also have a strict constraint on spintronic device writing speed. Spintronic memristive effects state that the device writing speed is roughly proportional to current magnitude because a fixed amount of the charge is required to pass through the device to change its state. Another consequence is that Joule heating during device writing is proportional to its operation speed. This is because Joule heating is proportional to charge multiplying voltage. For the requirement of a given amount of the charge to switch the device, the Joule heating is proportional to current, which is proportional to the switching speed.

4.4.1 Fastest Magnetization Switching with Minimum Joule Heating

Because a memristive device's behavior depends upon time-dependent current or voltage profile, in order to relieve memristive effects' constraint on device performance, we first optimize the electric voltage profile and magnetization dynamics to achieve the fastest magnetization speed with the smallest Joule heating.

We consider the magnetization switching mode in Equation (4.3) corresponding to an MTJ free layer element with a perpendicular uniaxial anisotropy energy $\varepsilon = (\sin^2 \theta)/2$. We optimize the current pulse shape to minimize magnetization switching time and Joule heat generation.

In an MTJ-based magnetic device, resistance depends upon the magnetization angle through Equation (4.1). Joule heating during magnetization switching is proportional to

$$\int\limits_0^\tau v^2(t)\left[\frac{G_{AP}+G_P}{2}\pm\frac{G_{AP}-G_P}{2}\cos\theta(t)\right]dt$$

where $v(t)$ is the voltage pulse magnitude, τ is the magnetization switching time, and $\pm$ corresponds to the switch from an antiparallel state to a parallel state, and a switch from a parallel state to an antiparallel state. We define an optimal stochastic control cost function:

$$J = \tau + k\int\limits_0^\tau v^2(t)\left[1+R\big[\theta(t)\big]\right]dt \tag{4.10}$$

where $R(\theta) = r\cos(\theta)$, with

$$r = \mp\frac{G_P - G_{AP}}{G_P + G_{AP}} = \mp\frac{TMR}{2+TMR}.$$

k in Equation (4.10) is a constraint parameter determining trade-offs between switching speed cost and heat generation cost. For a chosen k, optimizing the cost function (4.10) with magnetization dynamics (4.3) gives the solution for a minimum heat generation and a maximum switching speed. Varying control constraint parameter k gives the curve of the minimum heat generation versus the maximum magnetization switching speed.

Optimal stochastic control of Equations (4.3) and (4.10) can be obtained by solving the corresponding Hamilton-Jacobi-Bellman (HJB) equation [40]:

$$\inf_h\left\{\delta\frac{d^2J}{d\theta^2}+\delta ctg\theta\frac{dJ}{d\theta}+\sin\theta(v-\cos\theta)\frac{dJ}{d\theta}+1+k(1+R(\theta))h^2\right\}=0 \tag{4.11}$$

$$\left.\frac{dJ}{d\theta}\right|_{\theta=0}=0, \quad J(\theta_f)=0$$

where $J(\theta)$, $\tau(\theta)$ corresponds to cost and switching time for a magnetization starting from initial position θ. θ_f is the mean magnetization angles at the final state. The initial and final magnetization angles are determined by the equilibrium Boltzmann distribution at magnetization metastable states.

Functional minimization of the left side of Equation (4.11) with respect to the voltage pulse profile $v(t)$ gives the optimal control pulse shape:

$$v^* = -\frac{\sin\theta}{2k(1+R(\theta))}\frac{dJ}{d\theta} \tag{4.12}$$

Substituting Equation (4.12) into Equation (4.11) results in a closed equation for optimal control cost function:

$$\delta\frac{d^2J}{d\theta^2}+\delta ctg\theta\frac{dJ}{d\theta}-\frac{1}{4k(1+R(\theta))}\sin^2\theta\left(\frac{dJ}{d\theta}\right)^2-\sin\theta\cos\theta\frac{dJ}{d\theta}+1=0 \tag{4.13}$$

$$\frac{dJ}{d\theta}\bigg|_0=0,\quad J(\theta_f)=0$$

Equation (4.13) is a nonlinear differential equation that requires numerical solution in general. For a magnetic element with sufficient thermal stability at finite room temperature, $\delta \ll 1$, the asymptotic solution of Equation (4.13) can be constructed analytically:

$$J(\theta)=\int_{\theta_f}^{\theta}\frac{dJ}{d\theta}d\theta,$$

$$c_{1,2}=-2k(1+r)\left[1(+,-)\sqrt{1+1/k(1+r)}\right],$$

$$d_{1,2}=2k(1-r)\left[1(-,+)\sqrt{1+1/k(1-r)}\right]$$

$$\frac{dJ}{d\theta}=2k(1+r\cos(\theta))\frac{-\cos\theta-\sqrt{\cos^2\theta+1/(k(1+r\cos(\theta)))}}{\sin\theta} \tag{4.14}$$

$$+\frac{c_1c_2\left(1-e^{(c_1-c_2)\theta^2/(8k(1+r)\delta)}\right)}{\theta\left(c_2-c_1e^{(c_1-c_2)\theta^2/(8k(1+r)\delta)}\right)}-\frac{c_1}{\theta}$$

$$-\frac{d_1d_2\left(1-e^{(d_1-d_2)(\theta-\pi)^2/(8k(1-r)\delta)}\right)}{(\theta-\pi)\left(d_2-d_1e^{(d_1-d_2)(\theta-\pi)^2/(8k(1-r)\delta)}\right)}+\frac{d_1}{(\theta-\pi)}$$

With Equations (4.14) and (4.11), the average exit time can be obtained by solving the following linear differential equation:

$$\delta\frac{d^2\tau}{d\theta^2}+\delta ctg\theta\frac{d\tau}{d\theta}+(v^*-\cos\theta)\sin\theta\frac{d\tau}{d\theta}+1=0$$

$$\left.\frac{d\tau(\theta)}{d\theta}\right|_0=0,\quad \tau(\theta_f)=0$$

(4.15)

Equation (4.15) has an explicit analytical solution:

$$\tau(\theta)=\int_\theta^{\theta_f}\frac{dy}{\varphi(y)}\int_0^y\frac{\varphi(x)}{\delta}dx$$

$$\varphi(x)=\exp\left[\int_0^x\frac{\delta ctgz+(-\sin z(dJ(z)/dz)/(2k)-\cos z)\sin z}{\delta}dz\right]$$

(4.16)

Solutions $J(\theta)$ and $\tau(\theta)$ in Equations (4.14) and (4.16) and the optimal pulse $v(\theta)$ in Equation (4.12) provide all the information for the optimal stochastic control of magnetization switching with the maximum switching speed and minimum heat generation.

Figure 4.11 shows the minimum Joule heating versus the switching speed for two different thermal agitation strengths. For comparison, Joule heating versus magnetization switching speed for effective pulses with constant amplitude is also shown in the same figure. The figures are magnetization

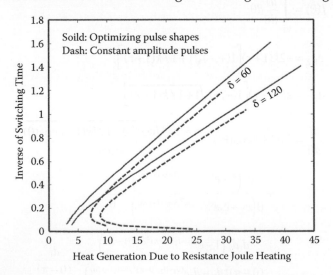

FIGURE 4.11
Normalized Joule heating versus normalized magnetization switching time for optimal pulses (solid) and constant amplitude pulses (dash) at two different thermal agitation strengths. The switching is from the antiparallel state to the parallel state. (From X. Wang and Y. Gu, *J. Appl. Phys.*, 113, 126106, 2013. With permission.)

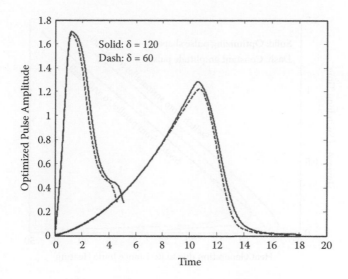

FIGURE 4.12
Optimal pulse shapes that minimize Joule heating energy for different magnetization switching times at two thermal agitation strengths. (From X. Wang and Y. Gu, *J. Appl. Phys.*, 113, 126106, 2013. With permission.)

switching from antiparallel states to parallel states. TMR is 100%. Significant heating reduction is obtained through pulse shape optimization, especially for a relatively long pulse width. The optimal pulse shapes for different switching speeds at two thermal agitation strengths are shown in Figure 4.12. Figure 4.11 shows an almost linear dependence of heat generation upon switching speed at relatively fast magnetization switching.

Figure 4.13 compares switching from the parallel state to the antiparallel state to switching from the antiparallel state to the parallel state. Pulse shape optimization is more effective for switching from the antiparallel state to the parallel state.

4.4.2 Minimum Heat Dissipation Intrinsic to Magnetization Reversal

Because of the memristive effects, spin torque device writing Joule heating is proportional to its operation speed. For a spintronic device based upon magnetization switching without electrons transporting across the device, the Joule heating is largely eliminated. This is the promise of low-power spintronic devices based upon electric or voltage-induced magnetization switching. However, even for devices without electronic transport across the device, nonequlibrium magnetization switching at finite temperature generates heat. It is of interest to understand dynamic effects in the heat generation process intrinsic to magnetization switching.

The minimum intrinsic magnetization switching heat generation is directly connected to the minimum heat requirement for information erasure. The

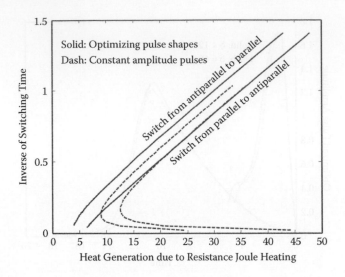

FIGURE 4.13
Normalized Joule heating versus normalized magnetization switching time for optimal pulses (solid) and constant amplitude pulses (dash) for different switching directions. (From X. Wang and Y. Gu, *J. Appl. Phys.*, 113, 126106, 2013. With permission.)

famous Landauer principle [41] states that the erasure of one bit of information during an operation at a thermal environment requires the release of heat (on average) of at least $\delta = \ln 2k_B T$. This is quite a small lower bound for heat generation. Unfortunately, it is well known that in order to reach the Landaeur lower bound, the dynamical system needs to move slowly, passing intermediate times through a sequence of equilibrium states. This directly conflicts with the fast magnetization switching requirement for spintronic devices. Thus, it is interesting to investigate the possibility of a dynamic magnetization switching that leads to such a minimum heat generation.

For the magnetization dynamics described by the stochastic LLG equation with energy $\varepsilon(\theta, t)$ (Equation (4.3)), the magnetization switching intrinsic heat generation can be written as [42]

$$Q = \int_0^\tau \left[\left(\frac{\partial \varepsilon}{\partial \theta} \right) \left(\frac{\partial \varepsilon}{\partial \theta} \right) - \frac{\delta}{\sin \theta} \left(\frac{1}{\sin \theta} \frac{\partial \varepsilon}{\partial \theta} \right) \right] dt \qquad (4.17)$$

where heat is normalized by the magnetization energy $H_k M_s V$. Solid-dash and solid curves in Figure 4.14 show the intrinsic magnetization switching heat generation versus the switching speed for pulses with constant amplitude and pulse shapes achieving minimum Joule heating, respectively. For magnetization switching intrinsic heat generation, the differences between constant amplitude pulses and pulses achieving minimum Joule heating are not significant. Linear dependence of heat generation upon magnetization speed is observed.

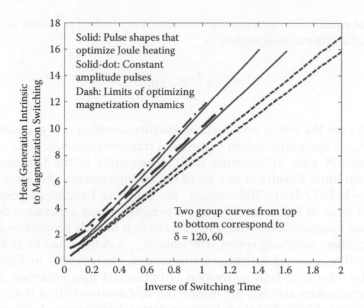

Solid: Pulse shapes that optimize Joule heating
Solid-dot: Constant amplitude pulses
Dash: Limits of optimizing magnetization dynamics

Two group curves from top to bottom correspond to δ = 120, 60

FIGURE 4.14
Normalized magnetization switching intrinsic heat generation versus normalized switching time for pulse shapes optimizing Joule heating (solid) and pulse shapes with constant amplitude (solid-dot). The dash curves are the minimum heat generation from optimized magnetization dynamics. (From X. Wang and Y. Gu, *J. Appl. Phys.*, 113, 126106, 2013. With permission.)

We could optimize the intrinsic magnetization switching heat generation following the same procedure as before using a cost function defined by $J = \tau + kQ$. However, in order to investigate the minimum heat generation intrinsic to magnetization switching, we do not confine to the specific magnetization energy format used so far. For example, in electric field-induced magnetization switching, the electric field could change magnetic anisotropy instead of the effective field. In order to pursue the minimum heat generation intrinsic to magnetization switching, instead of only optimizing pulse shape $h(t)$, we optimize the functional format of the time-dependent energy $\varepsilon(\theta, t)$ in Equation (4.1). In this case, the HJB equation is

$$\inf_{\varepsilon}\left\{ \delta\frac{d^2J}{d\theta^2} + \delta ctg\theta\frac{dJ}{d\theta} + \left(-\frac{\partial\varepsilon}{\partial\theta}\right)\frac{dJ}{d\theta} + 1 + kQ \right\} = 0$$

(4.18)

$$\left.\frac{dJ}{d\theta}\right|_{\theta_i} = 0, \quad J(\theta_f) = 0$$

This optimization problem has been solved for quite general stochastic dynamics [43]. It was shown in [43] that the solution is equivalent to the classical Monge-Kantorovich (MK) optimal mass transport solution [44]. The initial and final mass distributions in the MK solution correspond to the initial and final equilibrium probability distributions. Given a required switching time Δt, the optimal heat release is

$$Q_{opt} = \frac{K_{opt}}{\Delta t} - \delta \ln \frac{\rho_f}{\rho_i} \qquad (4.19)$$

where $\rho_{i,f}$ are the initial and final probability distributions at equilibrium states. K_{opt} is the minimization of the mass transport cost function.

The second term in Equation (4.19) corresponds to the Landauer heat generation limit. Erasure of one bit of digital information releases a heat of $-(-\ln 1 + 2 * \ln(1/2) \cdot (1/2)) \delta = \ln(2) \delta$, the same as Landauer's prediction. The first term in Equation (4.19) is the penalty of heat generation due to a finite time magnetization switching. This term is linearly proportional to the magnetization switching speed. The slope K_{opt} is determined by initial and final equilibrium magnetization distributions. Dash curves in Figure 4.14 show the minimum heat generation of optimized magnetization dynamics. Magnetization satisfies the Boltzmann distribution with a static energy $\varepsilon = \frac{1}{2} \sin^2 \theta$ at initial and final equilibrium states. The minimum heat generated during the magnetization switching increases linearly with magnetization switching speed. Its magnitude is proportional to magnetic stability energy $H_k M_s V$, which determines magnetization nonvolatility at finite temperature.

4.4.3 Effects of Polarization Direction and Precessional Magnetiation Switching

In previous MTJ examples, magnetization switching is achieved through a competition between spin torque excitation and magnetization relaxation. Because of the symmetric configuration of the perpendicular MTJ stack and spin polarization direction aligned to the perpendicular anisotropy direction, the magnetization precessional effect is suppressed and does not affect magnetization switching speed. However, magnetization switching utilizing gyromagnetic precession has been studied for quite a long time; see [45–47] for examples. For MTJ spin torque-induced magnetization switching, varying the polarization direction has been proposed to significantly reduce critical switching current and increase switching speed [48–51]. In this subsection, we will examine memristive effects in spin torque-induced magnetization switching by varying the polarization direction utilizing gyromagnetic precession.

Let's first review the difference between relaxation switching and precessional switching for magnetic field-induced magnetization switching. Figure 4.15(a) shows the directions of gyromagnetic torque and damping torque in the Landau-Lifshitz-Gilbert equation (4.2). For DC magnetic field-induced magnetization switching, the switching is accomplished through the magnetization relaxation term, as shown in Figure 4.15(b). When a constant

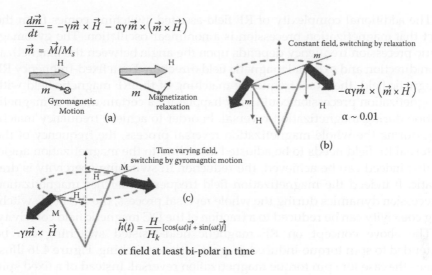

$$\frac{d\vec{m}}{dt} = -\gamma\vec{m}\times\vec{H} - \alpha\gamma\vec{m}\times\left(\vec{m}\times\vec{H}\right)$$

$$\vec{m} = \vec{M}/M_s$$

Gyromagnetic Motion

Magnetization relaxation

(a)

Constant field, switching by relaxation

$$-\alpha\gamma\vec{m}\times\left(\vec{m}\times\vec{H}\right)$$

$$\alpha \sim 0.01$$

(b)

Time varying field, switching by gyromagntic motion

(c)

$$\vec{h}(t) = \frac{H}{H_k}[\cos(\omega t)\vec{i} + \sin(\omega t)\vec{j}]$$

or field at least bi-polar in time

FIGURE 4.15
Magnetization switching through DC and AC magnetic fields. (From X. Wang et al., *CMOS Processors and Memories*, Springer, Berlin, 2010. With permission.)

magnetic field is applied to the opposite direction of the initial magnetization, the relaxation torque pulls magnetization toward the magnetic field direction (as shown in Figure 4.15(b)). However, due to the small damping parameter in ferromagnetic material, the magnitude of this relaxation torque usually is much smaller than the magnitude of the gyromagnetic torque for the same magnetic field amplitude. For a typical damping parameter around $0.005 \sim 0.01$ in ferromagnetic thin film, the magnitude of the relaxation torque is 100 to 200 times smaller than the magnitude of the gyromagnetic torque. Thus, if gyromagnetic torque can be used to directly switch magnetization, the switching magnetic field can be decreased significantly.

Figure 4.15(c) shows a configuration to switch magnetization through gyromagnetic torque. It can be seen that a magnetic field perpendicular to the initial magnetization direction provides a gyromagntic torque pulling magnetization away from the initial equilibrium condition. However, in order to successfully switch the magnetization, the magnetic field direction needs to be changed during the magnetization precession period as shown in Figure 4.15(c). If the magnetic field direction is constant, as in the DC switching case, the gyromagnetic torque directions in the first half period and second half period of the magnetization precession are exactly opposite. The averaged gyromagnetic torque on the magnetization vector will be zero during one period of magnetization precession. Because magnetization precession is in the GHz frequency range, the successful switching of magnetization through the AC magnetic field requires a magnetic field direction changing in the GHz radio frequency range.

The additional complexity of RF field-assisted switching comes from the fact that magnetization precession is a nonlinear oscillation. The gyromagnetic precession frequency depends upon the angle between the magnetization direction and external magnetic field direction. For a fixed-frequency RF magnetic field, the exact frequency matching of the RF magnetic field with magnetization precession could only happen for a certain angle of magnetization during magnetization reversal. In order to achieve frequency matching during the whole magnetization reversal process, the frequency of the external RF field needs to be adjusted according to the magnetization angle. If this indeed can be achieved, the reduction in switching coercivity is dramatic. If indeed the magnetization field frequency matches magnetization precession dynamics during the whole reversal process, the AC field switching coercivity can be reduced to a fraction of the DC magnetic field coercivity.

The above concept on RF magnetic field-assisted switching can be extended to spin torque-induced magnetization switching. Figure 4.16 illustrates the case for spin torque magnetization reversal. Instead of a fixed spin polarization direction, the spin polarization direction changes with time. If the direction of spin polarization can match magnetization precession to provide maximum spin torque during magnetization reversal, a significant reduction in switching current can be achieved. Figure 4.17 compares the switching current magnitude solution of the constant polarization direction to the optimal time-varying polarization direction for a thin-film element. The optimal spin polarization direction solution is obtained by a time-varying polarization direction that provides maximum spin torque during the whole magnetization reversal process. The magnetization trajectory and spin polarization direction evolutions are shown in the figure. Significant reduction in critical switching current is obtained.

Exact matching of the polarization direction to the magnetization direction during whole spin torque magnetization reversal is difficult to achieve for practical applications. However, there could be intermediate solutions between the

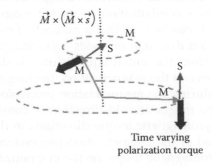

FIGURE 4.16
Reduction of current magnitude for current spin polarization direction changes with time, providing maximum spin torque during the reversal process. (From X. Wang et al., *CMOS Processors and Memories*, Springer, Berlin, 2010. With permission.)

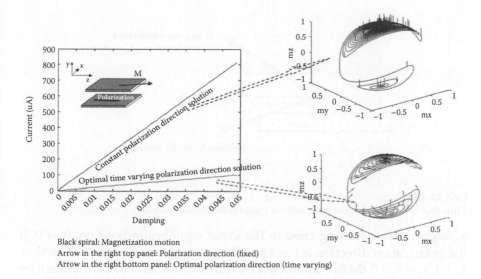

Black spiral: Magnetization motion
Arrow in the right top panel: Polarization direction (fixed)
Arrow in the right bottom panel: Optimal polarization direction (time varying)

FIGURE 4.17
Reduction in switching current by the optimal time-varying polarization direction matching the magnetization direction to provide maximum spin torque during the magnetization reversal process. (From X. Wang et al., *CMOS Processors and Memories*, Springer, Berlin, 2010. With permission.)

constant polarization direction and optimal time-varying polarization direction that may be easy to implement practically. Also, for the constant polarization direction, optimizing the polarization angle could be effective in reducing the critical switching current and increasing the switching speed.

Let's consider a thin-film element (Figure 4.18) with equilibrium in-plane magnetization pointing to a direction determined by shape anisotropy. The spin polarization can be at an arbitrary direction. The LLG equation with the spin torque term (4.1) in the spherical coordinate is

$$\frac{d\theta}{dt} = -\frac{1}{\alpha}(D_y - D_x)\sin\theta\sin\phi\cos\phi - \sin\theta\cos\theta(D_x\cos^2\phi + D_x\cos^2\phi - D_z)$$

$$+ h(\sin\theta n_z - \cos\theta\cos\phi n_x - \cos\theta\sin\phi n_y)$$

$$\frac{d\phi}{dt} = \frac{1}{\alpha}(D_x\cos^2\phi + D_y\sin^2\phi - D_z)\cos\theta - (D_y - D_x)\sin\phi\cos\phi$$

$$+ h(\sin\phi n_x - \cos\phi n_y)/\sin\theta$$

(4.20)

where $m_z = \cos\theta$, $m_x = \sin\theta\cos\phi$, and D_x, D_y, and D_z are shape anisotropies. For simplification, the thermal agitation term and nonadiabatic spin torque term are neglected. The benefit of the optimization polarization direction comes from spin torque terms $h(\sin\theta n_z - \cos\theta\cos\phi n_x - \cos\theta\sin\phi n_y)$ in the θ equation of (4.20) and $h(\sin\phi n_x - \cos\phi n_y)/\sin\theta$ in the ϕ equation of (4.20). For

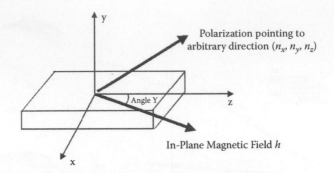

FIGURE 4.18

Thin-film magnetic element with arbitrary spin polarization direction.

a magnetization starting close to the initial equilibrium position, $\theta \to 0$. If the polarization direction is fixed parallel to the shape anisotropy direction, $n_z = 1, n_x = n_y = 0$, the θ component of the spin torque term will be very small for initial magnetization switching: $h(\sin\theta n_z - \cos\theta \cos\phi n_x - \cos\theta \sin\phi n_y) \to h\theta$. Thus, it is desirable to have a polarization not aligned to the shape anisotropy direction: $n_x, n_y \neq 0$.

However, it is important to realize is that this argument only holds for magnetization switching close to the initial equilibrium position. Once magnetization starts to rotate, the polarization direction away from shape anisotropy could hurt magnetization switching.

For ferromagnetic dynamics, gyromagnetic motion is much faster than relaxation due to the small damping parameter. Gyromagnetic motion conserves magnetic energy and has a constant energy level trajectory on the energy surface. Gyromagnetic motion and magnetization relaxation motion have well-separated timescales. The gyromagnetic motion timescale is in subnanoseconds. Depending upon damping, the magnetization relaxation timescale could be tens to hundreds times slower than the gyromagnetic timescale. For spin torque-induced magnetization switching, if spin torque excitation cannot swing magnetization away from the initial equilibrium position at an ultra-fast gyromagnetic motion timescale, magnetization will start to rotate around a constant energy level and magnetization switching will be controlled by magnetization relaxation. In this case, the equation of motion (4.20) could be averaged around the constant energy level. It was shown both theoretically and experimentally in [52] that for the averaged magnetization dynamics, only the n_z component of the spin torque contributes to magnetization switching. This is consistent with well-known results that the critical switching current for spin torque-induced magnetization is $1/n_z$. Thus, in order to utilize the angular effect of spin polarization, the magnetization switching must happen at a gyromagnetic timescale with well-controlled current pulse duration.

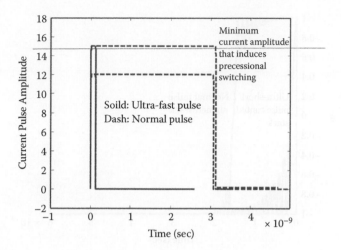

FIGURE 4.19
Ultra-fast pulse for precessional switching and normal pulse for relaxation switching.

In order to illustrate the above concepts, we consider an example of two pulse-exciting MTJ stacks. The first pulse is ultra-fast (~100 ps), and the second pulse has a normal duration (~nanosecond), as shown in Figure 4.19. The pulses all have very quick rise times in picoseconds. The first MTJ stack has a fixed polarization direction pointing out of a thin-film plane and 60° canted away from the shape anisotropy direction. This is called a canted stack. The second MTJ stack has a polarization aligned to the thin-film shape anisotropy direction (z-direction). This is called an aligned MTJ stack. The thin-film element of the MTJ has demagnetization $D_x = 0.69$, $D_y = 11.68$, and $D_z = 0.29$. The magnetization saturation is $M_s = 1090$ emu/cc. The damping parameter is chosen to be $\alpha = 0.01$.

When the ultra-fast current pulse amplitude reaches a sufficient magnitude, it can switch the canted MTJ stack very fast. This is shown in Figure 4.20. This current amplitude is defined as a critical precessional switching current. The ultra-fast pulse with critical precessional switching magnitude cannot switch an aligned MTJ stack. For an aligned MTJ stack, a normal pulse with a duration in the nanosecond region is required to switch its magnetization. This is also shown in Figure 4.20. Here the amplitude of the normal pulse is the same as the amplitude of the ultra-fast pulse (above the critical precessional switching amplitude).

Now if we reduce the current amplitude below the critical precessional switching current, as shown in Figure 4.21, an ultra-fast pulse cannot switch canted MTJ. It does not provide a torque sufficient to swing magnetization out of the initial equilibrium condition. With the same amplitude below the critical precessional switching, a normal pulse with a much longer duration also cannot switch canted MTJ. This is shown in Figure 4.21. As discussed before, when initial torque is not sufficient to swing magnetization out of the initial

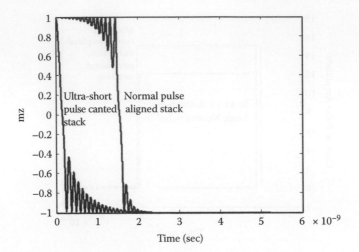

FIGURE 4.20
Magnetization switching for pulse amplitude reaching critical precessional switching current.

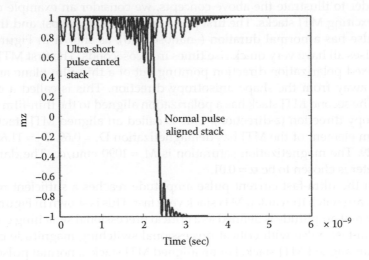

FIGURE 4.21
Magnetization switching for pulse amplitude below critical precessional switching current.

equilibrium condition, magnetization will rotate around a constant energy surface. The average effect sets in and magnetization cannot be switched. On the other hand, for current with an amplitude below the critical precessinal switching, a normal pulse can switch the aligned stack, although the switching time is longer than in the case of a normal pulse with an amplitude above the critical precessional current. This is shown in Figure 4.21.

Figures 4.22 and 4.23 illustrate this more clearly through three-dimensional magnetization trajectories. As shown in Figure 4.22, for current amplitude

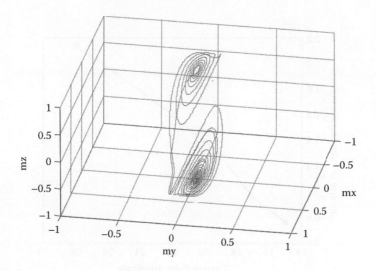

FIGURE 4.22
See color insert. Magnetization trajectory for pulse amplitude reaching critical precessional switching current. Red, ultra-short pulse; black, normal pulse.

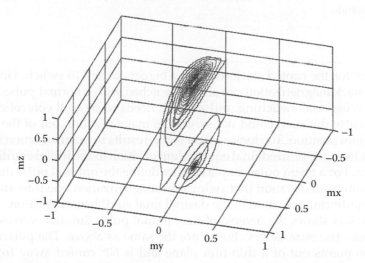

FIGURE 4.23
See color insert. Magnetization trajectory for pulse amplitude below critical precessional switching current. Red, ultra-short pulse; black, normal pulse.

above critical precessional switching current, the initial torque of the canted MTJ stack quickly swings magnetization out of its initial equilibrium position, while magnetization rotates many circles for the aligned MTJ stack to switch. When current amplitude is below the critical precessional switching current, magnetization in the canted stack starts to rotate around the initial equilibrium position. This is shown in Figure 4.23. The averaged magnetization

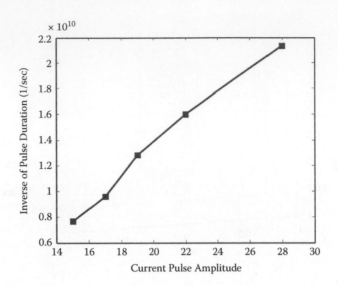

FIGURE 4.24
Inverse of precessional switching optimal (minimum) current pulse duration versus current pulse amplitude.

dynamics for the canted stack requires a bigger current to switch. However, aligned stack magnetization can still be switched with a normal pulse.

For precessional switching, pulse width needs to be well controlled. Too short a pulse duration could not swing the magnetization out of the initial equilibrium position. Too long a pulse width results in magnetization swinging back from the desired final equilibrium position to the initial equilibrium condition. For a given pulse amplitude, we define the optimal pulse duration as the minimum duration that swings the magnetization one time from the initial equilibrium position to the desired final equilibrium position.

Figure 4.24 shows the inverse of the optimal pulse duration versus pulse amplitude. The parameters chosen are the same as above. The polarization direction points out of a thin-film plane and is 60° canted away from the shape anisotropy direction. Figure 4.24 shows the linear dependence of the inverse of the optimal pulse duration versus current amplitude. This means that for spin torque-induced ultra-fast precessional magnetization switching, a minimum amount of the charge is required to pass through the stack to switch magnetization. This is the same as the memristive effects in the previous relaxation magnetization switching.

The above analysis could be further extended to the time-varying spin polarization direction. To illustrate memristive effects in the time-varying spin polarization direction case, we consider Equation (4.20) with a simplified case of $D_x = D_y$:

$$\frac{d\theta}{dt} = -\sin\theta\cos\theta(D_x - D_z)$$

$$+ h(\sin\theta n_z(t) - \cos\theta\cos\phi n_x(t) - \cos\theta\sin\phi n_y(t)) \qquad (4.21)$$

$$\frac{d\phi}{dt} = \frac{1}{\alpha}(D_x - D_z)\cos\theta + h(\sin\phi n_x(t) - \cos\phi n_y(t))/\sin\theta$$

where $[n_z(t), n_x(t), n_y(t)]$ is the time-dependent polarization direction. The optimal time-dependent polarization direction corresponds to

$$\sup_{\left[n_x(t), n_x(t), n_y(t)\right]} \left\{\sin\theta n_z(t) - \cos\theta\cos\phi n_x(t) - \cos\theta\sin\phi n_y(t)\right\} \qquad (4.22)$$

Notice {$\sin\theta$, $-\cos\theta\cos\phi$, $-\cos\theta\sin\phi$} is a unit vector perpendicular to magnetization direction {$\cos\theta$, $\sin\theta\cos\phi$, $\sin\sin\phi$}, and $[n_z(t), n_x(t), n_y(t)]$ is a unit vector in the direction of polarization. The sup in (4.22) is obtained by aligning the polarization vector direction to the vector, and the vector direction perpendicular to the magnetization direction. In this case, magnetization switching is simplified to

$$\frac{d\theta}{dt} = -\alpha\sin\theta\cos\theta(D_x - D_z) + h \qquad (4.23)$$

The minimum pulse duration to bring magnetization from the initial magnetization state to the critical magnetization switching position is

$$T = \int_0^{\pi/2} \frac{d\theta}{-\sin\theta\cos\theta(D_x - D_z) + h} \qquad (4.24)$$

Figure 4.25 shows the inverse of pulse duration versus current pulse magnitude. Again, memristive effects are observed.

4.5 Spintronic Device Reading Operation Memristive Effects

Memristive effects are universal for spin torque spintronic devices at the timescale that explicitly involves the interactions between magnetization dynamics and electronic charge transport. In this section, I show that memristive effects not only affect magnetization switching at the short-time dynamic write region, but also affect thermal stability at the long-time read process.

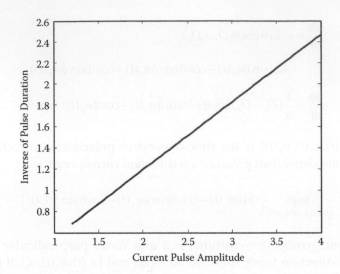

FIGURE 4.25
Inverse of switching current pulse duration versus current pulse amplitude for time-optimized varying polarization angle.

For the read process with a constant amplitude pulse, magnetization switching probability that determines the error rate can be written as

$$P = 1 - e^{-\gamma T^{\,*}} \tag{4.25}$$

where P is switching probability, T is pulse duration time, and γ is transition rate. The transition rate follows the Neel-Arrenhius formula [53]:

$$\gamma = f_0 e^{-\delta E/k_B T} \tag{4.26}$$

where f_0 is the attempt frequency that could be temperature and excitation dependent. $f_0 = f_0(h,T)$, where h denotes the external excitation amplitude.

For the magnetic thin-film element model (4.3) with normalized energy corresponding to the free layer perpendicular anisotropy case, the energy barrier during a constant amplitude read pulse excitation is

$$\frac{\delta E}{(D_x - D_z)M_s^2} = \left(1 - \frac{I}{I_c}\right)^2 \approx 1 - 2\frac{I}{I_c} \tag{4.27}$$

where I_c is the critical switching current and the Taylor expansion has been used for small sensing current amplitude. The square dependence of the reversal barrier on the switching current is due to the rotational symmetry of the magnetic element [54]. Based on the above equation, the probability of magnetization switching during one pulse excitation is

$$1 - \exp\left(-Tf_0 e^{-\frac{(D_x - D_z)M_s^2}{k_B T}\left(1 - 2\frac{I}{I_c}\right)}\right) \tag{4.28}$$

For a given switching probability or error rate, the current pulse amplitude is proportional to the logarithmic of the pulse width. This scaling behavior is quite different from memristive scaling, where current pulse amplitude is proportional to the inverse of the pulse width.

However, it has been shown in [55–57] that the above scaling only holds for relatively long pulse durations. A "scale dilemma" results if we insist on using the above formula to describe short current pulse duration (~GHz) read disturbance. The reason is that detailed magnetization dynamics must be included to properly describe magnetization nonvolatility as read excitation frequency approaches the magnetization dynamical timescale. The above time-dependent energy barrier approach neglects the fast magnetization dynamic and is not sufficient to describe magnetization switching probability for ultra-short pulse duration excitation.

Under the assumption that the spin torque excitation magnitude is well below the critical switching value, and based upon the stochastic Landau-Lifshitz-Gilbert equation with a spin torque term, the switching barrier reduction due to spin torque current excitation can be derived as [55, 56]

$$\delta E = \min_{t_c} \int_{-\infty}^{\infty} \chi(t) \cdot \beta(t - t_c) dt \tag{4.29}$$

where $\beta(t)$ is the normalized spin torque proportional to polarized current magnitude, and $\chi(t)$ is the magnetization logarithmic susceptibility. The magnetization logarithmic susceptibility is defined as the ratio of reversal barrier reduction to external spin torque current excitation magnitude. t_c in Equation (4.29) is the best time for the magnetization reversal to happen. Periodic external forcing lifts the time degeneracy of the escape path. It synchronizes optimal escape trajectories, one per period, to minimize the activation energy of escape.

Magnetization logarithmic susceptibility $\chi(t)$ can be obtained from optimal reversal path $z_0(t)$, which is the minimization of functional action of the Landau-Lifshitz-Gilbert (LLG) equation with the spin torque term [55–57]. For a magnetic element with magnetization energy (4.3) and polarization pointing to the easy anisotropy direction,

$$\chi(t) = \left[D_x - D_z\right]\cos\left\{\tan^{-1}\left[e^{\alpha(D_x - D_z)t}\right]\right\}\left\{1 - \cos\left\{\tan^{-1}\left[e^{\alpha(D_x - D_z)t}\right]\right\}^2\right\} \tag{4.30}$$

Although the term *switching barrier reduction* has been used in describing the logarithmic susceptibility concept, the "barrier reduction" obtained through logarithmic susceptibility considers detailed magnetization dynamics. The LS switching barrier is not a static energy barrier based upon magnetization energy surface analysis. In the magnetization LS approach, through considering detailed magnetization dynamics, the dynamics timescale at high frequency (~GHz) is retained for the long-time thermal magnetization reversal process. Thus, long-time thermal magnetization reversal dependence upon excitation frequency (up to GHz) is properly characterized.

As pulse duration decreases, the pulse width effect on switching probability increases dramatically. This is because pulse duration directly affects thermal switching barrier reduction through magnetization logarithmic susceptibility high-frequency components. Considering magnetization dynamics at high frequency, Equation (4.28) needs to be revised to

$$1-\exp\left(-Tf_0e^{-\frac{(D_x-D_z)M_s^2}{k_BT}\left(1-r(T)\frac{I}{I_c}\right)}\right) \qquad (4.31)$$

where $r(T)$ is the dynamic switching barrier reduction factor calculated from logarithmic susceptibility. Figure 4.26 shows the dynamic switching barrier reduction factor as a function of pulse duration. It becomes more and more significant as pulse duration decreases.

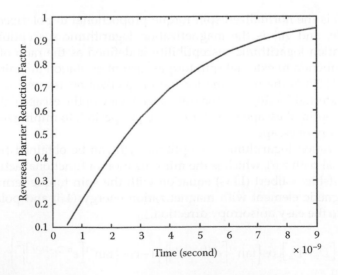

FIGURE 4.26
Dynamic switching barrier reduction factor as a function of pulse duration. (From Z. Wang, et al., *Appl. Phys. Lett.* 103, 142419, 2013. With permission.)

For a given switching probability, to the leading order, switching current amplitude and switching pulse duration can be related as

$$I \sim I_c / r(T) \tag{4.32}$$

Figure 4.27 shows the inverse of current pulse duration versus current pulse amplitude for constant error rate or switching probability. When dynamic effects are properly included, long-time thermal reversal shows obvious memristive effects.

Experiment measurements have been performed to verify the prediction of logarithmic susceptibility and the dynamic process in the spin torque random access memory read process [54, 58]. Figure 4.28 shows an example of measured current pulse duration versus current pulse amplitude [58]. The experiment confirms theoretical prediction of magnetization logarithmic susceptibility. Memristive behaviors are obvious in both experimental and theoretical models.

It should be pointed out that read operation memristive effects could have significant impacts on spin torque random access memory device operation and design. Due to the significant reduction of programming current for spin torque random access memory with improvement of magnetic tunneling junction material and structure, the sense current margin for STT-MRAM design has been greatly tightened. Spintronic read memristive effects provide dramatic reduction of read disturbance through pulse width control,

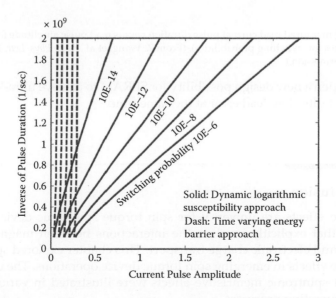

FIGURE 4.27
Inverse of read current pulse duration versus read pulse amplitude for different switching probabilities (or bit error rates). (From Z. Wang, et al., *Appl. Phys. Lett.* 103, 142419, 2013. With permission.)

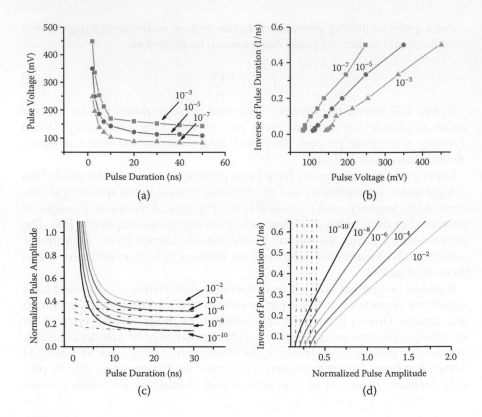

FIGURE 4.28

Modeled and measured read current pulse duration versus read pulse amplitude for different read error rates (or switching probabilities). (From Z. Wang, et al., *Appl. Phys. Lett.* 103, 142419, 2013. With permission.)

and thus open a new design possibility for SPRAM to achieve ultra-low write current and ultra-low read error at the same time.

4.6 Conclusion

Memristive effects are universal for spin torque spintronic devices at the timescale that explicitly involves the interactions between magnetization dynamics and electronic charge transport. This chapter explored spintronic memristive effects in emerging spin torque device operations. The universal features of spintronic memristive effects were illustrated in various write and read operating processes.

Although memristive effects could provide many exciting new application possibilities for spintronic devices, they also have quite strict constraints on spintronic device write operations, such as writing speed and heat dissipation. We studied spintronics device write optimization in light of its effects.

Memristive effects not only affect spintronic device write operation at the short-time dynamic magnetization reversal region, but also affect device read operations at the long-time thermal stability region. Experiment measurement and theoretical calculation reveal obvious memristive behaviors for short-pulse read operations. Dramatic reduction in read disturb through pulse width control provides a new design possibility for ultra-fast low-current STT-MRAM.

References

1. L. Berger, *Phys. Rev. B* 54, 9353, 1996.
2. J. C. Slonczewski, *J. Magn. Magn. Mater.* 159, L1, 1996.
3. W. F. Brown Jr., *Phys. Rev.* 130, 1677, 1963. I. N. Krivorotov, N. C. Emley, A. G. F. Garcia, J. C. Sankey, S. I. Kiselev, D. C. Ralph, and R. A. Buhrman, *Phys. Rev. Lett.* 93, 166603, 2004. Z. Li and S. Zhang, *Phys. Rev. B* 69, 134416, 2004. D. M. Apalkov and P. B. Visscher, *Phys. Rev. B* 72, 180405(R), 2005. J. L. Garcia-Palacios and F. J. Lazaro, *Phys. Rev B* 58, 14937, 1998.
4. I. Theodonis, N. Kioussis, A. Kalitsov, M. Chshiev, and W. H. Butler, *Phys. Rev. Lett.* 97, 237205, 2006.
5. S. Salahuddin, D. Datta, P. Srivastava, and S. Datta, at *IEEE Electronic Device Meeting*, 2007.
6. J. Xiao and G. E. Bauer, *Phys. Rev. B* 77, 224419, 2008.
7. C. Heiliger and M. D. Stitles, *Phys. Rev. Lett.* 100, 186805, 2008.
8. A. Rebei, W. N. G. Hitchon, and G. J. Parker, *Phys. Rev. B* 72, 064408, 2005.
9. A. Manchon, N. Ryzhanova, N. Strelkov, A. Vedyayev, M. Chshiev, and B. Dieny, *J. Phys. Condens. Matter* 20, 145208, 2008.
10. P. M. Haney, R. A. Duine, A. S. Nunezc, and A. H. MacDonald, *J. Magn. Magn. Mater.* 320, 1300, 2008.
11. X. Wang, W. Zhu, and D. Dimitrov, *Phys. Rev. B* 79, 104408, 2009.
12. X. Wang, W. Zhu, and D. Dimitrov, *Phys. Rev. B* 78, 024417, 2008.
13. L. O. Chu, *IEEE Trans. Circuit Theory* 18, 507, 1971.
14. X. Wang, Y. Chen, H. Xi, and D. Dimitrov, *IEEE Electron Device Lett.* 30, 294, 2009.
15. G. Tatara and H. Kohno, *Phys. Rev. Lett.* 92, 086601, 2004.
16. R. A. Duine, A. S. Nunez, and A. H. MacDonald, *Phys. Rev. Lett.* 98, 056601, 2007.
17. D. B. Strukov, G. S. Snider, D. R. Stewart, and R. S. Williams, *Nature* 453, 80, 2008.
18. M. Di Ventra and Y. Pershin, *Mater. Today* 14, 584, 2011.
19. X. Wang and Y. Chen, in *Design, Automation and Test Conference in Europe*, 2010, IEEE 10.1109/DATE.2010.5457118.
20. X. Lou, Z. Gao, D. V. Dimitrov, and M. X. Tang, *Appl. Phys. Lett.* 93, 242502, 2008.
21. E. Linn, R. Rosezin, C. Kugeler, and R. Waser, *Nat. Mater.* 9, 403, 2010.
22. W. Zhao, S. Chaudhuri, C. Accoto, J.-O. Klein, C. Chappert, and P. Mazoyer, *IEEE Trans. Nanotechnol.* 11, 907, 2012.

23. G. Snider, *Appl. Phys. A* 80, 1165, 2009.
24. D. Strukov and K. Likharev, *Nanotechnology*, 16, 888, 2005.
25. D. Strukov and K. Likharev, *IEEE Trans. Nanotechnol.* 6, 696, 2007.
26. S. Matsunaga, J. Hayakawa, S. Ikeda, K. Miura, H. Hasegawa, T. Endoh, H. Ohno, and T. Hanyu, *Appl. Phys. Express* 1, 091301, 2008.
27. X. Lou, D. Dimitrov, and S. Xue, U.S. Patent 7855911, 2010.
28. S. Parkin, M. Hayashi, and L. Thomas, *Science* 320, 190, 2008
29. X. Wang, Y. Chen, Y. Gu, and H. Li, *IEEE Device Lett.* 31, 20, 2010.
30. G. Snider, *Nanotechnology*, 18, 1, 2007.
31. Y. Pershin and M. Ventra, *Phys. Rev. E* 84, 046703, 2011.
32. F. Merrikh-Bayat, S. Shouraki, and F. Merrikh-Bayat, in *2011 Eighth International Conference on Fuzzy Systems and Knowledge Discovery (FSKD)*, 2011, 0.1109/FSKD.2011.6019541.
33. M. Klimo, O. Such, in *Fourth International Conference on Future Computational Technologies and Applications*, 2012, arXiv, 1110.2047.
34. T. Driscoll, J. Quinn, S. Klein, H. Kim, B. Kim, Y. Pershin, M. Ventra, and D. Basov, *Appl. Phys. Lett.*, 97, 093502, 2010.
35. M. Itoh and L. Chua, *Int. J. Bifurcation Chaos*, 18, 3183, 2008.
36. M. Sharad, C. Augustine, G. Panagopoulos, and K. Roy, arXiv:1206.3227S.
37. S. Parkin, at *Saudi International Nanotechnology Conference*, 2010.
38. X. Wang, W. Zhu, S. Markus, and D. Dimitrov, *IEEE Trans. Magn.* 45, 2038, 2009.
39. X. Wang, Y. Chen, and T. Zhang, *CMOS Processors and Memories*, Springer, Berlin, 2010.
40. M. Bardi and I. Capuzzo-Dolcetta, *Optimal Control and Viscosity Solution of Hamilton-Jacobi-Bellman Equations*, Birkhauser, 2008.
41. R. Landauer, *IBM J. Res. Dev.* 5, 183, 1961.
42. K. Sekimoto, *Prog. Theor. Phys. Suppl.* 180, 17, 1998.
43. E. Aurell, C. Mejía-Monasterio, and P. Muratore-Ginanneschi, *Phys. Rev. Lett.* 106, 250601, 2011.
44. G. Monge, *Histoire de l'Academie Royale des Sciences*, Annee, 1781. Imprimerie Royale, Paris, 666, 1784. L. Kantorovich, *Acad. Sci. URSS*, 37, 199, 1942.
45. J. Zhu, at *MMM 2005 Conference*, Paper CC-12. J. Zhu, X. Zhu, and Y. Tang, at *TMRC 2007 Conference*, paper B6. J. Zhu, X. Zhu, and Y. Tang, *IEEE. Trans. Magn.* 44, 125, 2008.
46. Z. Z. Sun and X. R. Wang, *Phys. Rev. B* 73, 092416, 2006.
47. L. He and W. D. Doyle, *IEEE Trans. Magn.* 30, 4086. W. K. Hiebert, A. Stankiewicz, and M. R. Freeman, *Phys. Rev. Lett.* 79, 1134, 1997. T. M. Crawford, T. J. Silva, C. W. Teplin, and C. T. Rogers, *Appl. Phys. Lett.* 74, 3386, 1999. Y. Acremann, C. H. Back, M. Buess, O. Portmann, A. Vaterlaus, D. Pescia, and H. Melchior, *Science* 290, 492, 2000. C. Thirion, W. Wernsdorfer, and D. Mailly, *Nat. Mater.* 2, 524, 2003. D. Xiao, M. Tsoi, and Q. Niu, *J. Appl. Phys.* 99, 013903, 2006.
48. H. W. Schumacher, C. Chappert, P. Crozat, R. C. Sousa, P. P. Freitas, J. Miltat, J. Fassbender, and B. Hillebrands, *Phys. Rev. Lett.* 90, 017201, 2003.
49. A. D. Kent, B. Zyilmaz, and E. del Barco, *Appl. Phys. Lett.* 84, 3897, 2004.
50. K. Rivkin and J. B. Kettersen, *Appl. Phys. Lett.* 89, 252507, 2006.
51. X. R. Wang and Z. Z. Sun, *Phys. Rev. Lett.* 98, 077201, 2007.
52. X. Wang, W. Zhu, Z. Gao, H. Xi, D. Dimitrov, Angular dependent spin torque, *J. Appl. Phys.*, 105, 07D103, 2009.

53. L. Néel, *Ann. Géophys.* 5, 99, 1949.
54. X. Wang, W. Zhu, H. Xi, and D. Dimitrov, *Appl. Phys. Lett.* 93, 102508, 2008.
55. X. Wang, W. Zhu, H. Xi, Z. Gao, and D. Dimitrov, *Appl. Phys. Lett.* 93, 182506, 2008.
56. X. Wang and P. Ryan, *J. Appl. Phys.* 93, 182506, 2010.
57. X. Wang, *SPIN World Sci.* 2, 1240009, 2012.
58. Z. Wang, X. Wang, H. Gan, D. Jun, K. Satoh, T. Lin, Y. Zhou, J. Zhang, Y. Huai, Y. Chang, and T. Wu, *Appl. Phys. Lett.* 103, 142419, 2013.
59. X. Wang and Y. Gu, *J. Appl. Phys.*, 113, 126106, 2013.

5

Magnetic Insulator-Based Spintronics: Spin Pumping, Magnetic Proximity, Spin Hall, and Spin Seebeck Effects on Yttrium Iron Garnet Thin Films

Yiyan Sun, Zihui Wang, and Lei Lu

Department of Physics, Colorado State University, Fort Collins, Colorado

CONTENTS

Overview

The year 2010 witnessed two significant discoveries in the field of spintronics: (1) transfer of electrical signals with magnetic insulators[1] and (2) spin Seebeck effects in magnetic insulators.[2,3] The magnetic insulators used in both discoveries were magnetic garnets, which Dr. Kittel referred to as the fruit fly of magnetism about 50 years ago.[4] For (1), an initial electrical signal excites a spin wave in one end of a YIG film strip via the spin Hall effect; as the spin wave travels to the other end of the YIG strip, it is converted to an electrical signal via the inverse spin Hall effect. For (2), one establishes a temperature gradient along a YIG or $LaY_3Fe_5O_{12}$ film strip, and the latter produces a difference between the chemical potentials of spin-up and spin-down electrons in the film strip. This potential difference, also called spin voltage, can generate a spin current in a normal metal layer deposited on the end of the film strip. Although several features related to these effects remain unexplained, the effects clearly demonstrate that one can transmit spin currents into and out of magnetic garnets as well as use magnetic garnets to generate spin currents. This demonstration opens a new paradigm in the discipline of spintronics and opens the door to a new class of spintronic devices that make use of either pure YIG or substituted-YIG materials.[5,6]

It should be noted that thanks to the absence of Fe^{2+} ions, magnetic garnets generally have slower ferromagnetic relaxation than other ferrimagnetic materials.[7] Moreover, in the family of magnetic garnets, YIG and substituted-YIG have the lowest relaxation rates. Indeed, YIG materials have a lower relaxation rate than any other magnetic materials, with an intrinsic damping constant α of about 3×10^{-5} only. Because of this extremely small damping, YIG materials have found rather broad current and potential applications in microwave devices.[8–10] YIG sphere-based oscillators and filters, for example, are core devices in many microwave generators and analyzers.

The significance of new YIG-based spintronic devices originates from two features of YIG materials: (1) extremely small damping, as mentioned above, and (2) electrically insulating. The intrinsic damping constant α indicated above for YIG is two orders of magnitude smaller than that in transition metals, which are the materials of choice in current spintronic devices. This small damping characteristic is highly desirable for many applications, including spin-wave-based logic operations and generation of large spin currents. Insulator-based spintronic devices are particularly attractive from an engineering point of view. They not only require substantially lower power, but also involve no issues that are intrinsically associated with charge currents.

Organization

The chapter is organized as follows. Section 5.1 gives an introduction to YIG properties and the magnetic insulator-based spintronic effects. Section 5.2 demonstrates the experimental results on the spin pumping efficiency of nanometer-thick YIG. Section 5.3 reviews the enhanced damping in nano YIG/Pt that originated from the magnetic proximity effect. Sections 5.4 and 5.5 discuss the spin Hall and spin Seebeck effects observed in YIG/Pt and their influences on magnetic relaxation. Section 5.6 is the summary for this chapter.

5.1 Introduction

5.1.1 Crystal Structure and Magnetic Properties of YIG

YIG has been widely applied in microwave and magneto-optic industries since it was discovered by Bertaut and Forrat in 1956. As one synthetic garnet, YIG has the lowest magnetic losses at microwave frequency. It is also treated as the prototype material in magnetic research because of its cubic single-crystal structure, soft magnetism, low anisotropy, and high Verdet constant.

5.1.1.1 Crystal Structure of YIG

YIG has a body-centered cubic (BCC) structure with a lattice parameter of 12.376 ± 0.004 Å. A single YIG unit cell contains 24 yttrium, 40 iron, and

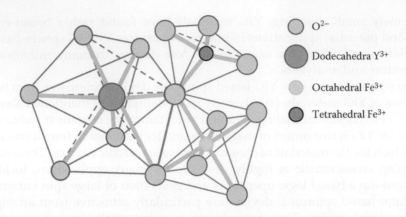

FIGURE 5.1
See color insert. Schematic diagram of three polyhedron sites in YIG crystal.

96 oxygen, for a total of 160 atoms. The unit cell can be separated into eight identical formula cells. In each formula cell, there are three types of coordination to oxygen, eightfold, sixfold, and fourfold, which correspond to three polyhedrons: dodecahedron, octahedron, and tetrahedron, respectively. The cation (Y^{3+} or Fe^{3+}) is surrounded by O^{2-} in the center of the polyhedrons. All the Y^{3+} ions are located in the dodecahedra sites, while Fe^{3+} ions occupy either tetrahedral or octahedral sites. The rigid spheres models of the three sites are shown in Figure 5.1.

To realize the ions arrangement in the YIG unit cell, one can simplify the structure by breaking down the unit cell into eight BCC formula cells. Figure 5.2 shows the structures with octahedral Fe ions only. Figure 5.2(a) shows the whole unit cell, and Figure 5.2(b) shows the formula cell, which is 1/8 of the unit cell. One can see the formula cell is a standard BCC structure. Because all the formula cells are identical, the YIG structure can be simply

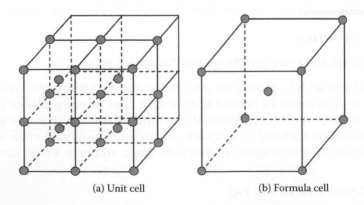

(a) Unit cell   (b) Formula cell

FIGURE 5.2
Octahedral Fe^{3+} in a unit cell (a) and a formula cell of YIG (b).

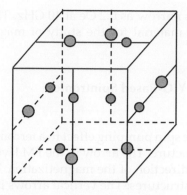

FIGURE 5.3
Tetrahedral Fe^{3+} and dodecahedral Y^{3+} in a formula cell of YIG.

learned within a single formula cell. Figure 5.3 shows both the tetrahedral Fe and dodecahedral Y ions in one formula cell. They just lay on the mid-lines of the cubic faces. There is only one single pair of Y and tetrahedral Fe ions on each line. For one pair, the two ions are located away from each other with the separation of a half-formula cell length.

5.1.1.2 Magnetic Properties of YIG

YIG is a ferrimagnetic material due to the antiparallel arrangement of Fe^{3+} spins in different sites. Its ferrimagnetism originates from the superexchange interactions between tetrahedral and octahedral Fe ions, mediated by the oxygen ions. The superexchange interaction model, originally proposed by Kramers in 1934, provides the explanation of antiferromagnetism between the two nearest equal valence cations through a nonmagnetic anion. The strength of the interaction is dependent on the angle of the magnetic ion–nonmagnetic ion–magnetic ion bond. The interaction is strongest when the angle is 180°. For YIG, the bond is tetrahedral Fe^{+3}–O^{2-}–octahedral Fe^{3+}, and the angle is 126.6°. The superexchange interaction results in the antiparallel arrangement between tetrahedral and octahedral Fe ions. In one unit cell, there are 24 Fe^{3+} occupying the tetrahedral site and 16 Fe^{3+} in the octahedral site. Each Fe^{3+} has the magnetic moment of five Bohr magnetons (5 μ_B), and therefore a unit cell of YIG has a 40 μ_B net magnetic moment. This is the theoretical value at the temperature of 0K. At room temperature, however, the standard value of $4\pi M_S$ is 1750 G. Overall, YIG is the ferrimagnetic material with its Curie temperature of 559K. Both the static and dynamic magnetic properties of YIG are similar to those of ferromagnetic materials. For example, YIG has a magnetic hysteresis property and its coercivity is as low as 50 Oe. The magnetization dynamics study shows YIG has extremely low intrinsic damping of 3×10^{-5}, which is two orders of magnitude smaller than in ferromagnetic metals. The ferromagnetic resonance linewidth in the single-crystal YIG is as narrow as 0.2 Oe at 10 GHz. These properties make YIG a good candidate material for the study of magnetic insulator-based spintronics.

5.1.2 Introduction to YIG-Based Spintronics

5.1.2.1 Spin Pumping

Figure 5.4 illustrates the spin pumping effect in a ferromagnetic (FM)/normal metal (NM) bilayer structure. The arrow in the FM layer shows a unit vector $\hat{m}$, which denotes the direction of the magnetization. The bottom diagrams show simplified band structures. The vertical arrows indicate the directions of electron magnetic moments. When the magnetization in the FM layer is static, the electrons in the two layers share the same Fermi level. When the magnetization is excited to precess, the FM layer passes a certain net angular momentum to the NM layer. The net effects are a difference between the chemical potentials of spin-up electrons ($\mu_\uparrow$) and spin-down electrons ($\mu_\downarrow$) and a spin current in the NM layer. This effect is called spin pumping.[11,12] Note that the small spheres in the NM layer represent electrons, and the arrow through each sphere indicates the direction of the electron magnetic moment. Note also that the chemical potential difference $\mu = \mu_\uparrow - \mu_\downarrow$ is often called a spin voltage or a spin accumulation.

One often defines a spin current density vector as

$$J_s = \frac{\hbar}{4\pi} \mathrm{Re}(g_{\uparrow\downarrow}) \hat{m} \times \frac{\partial \hat{m}}{\partial t} \qquad (5.1)$$

which denotes the spin mixing conductance across the FM/NM interface. This density is indeed the net angular momentum transferred from the spins in the FM layer to the electrons in the NM layer through a unit interface area within a unit time. This momentum transfer results in an enhancement in the damping constant α in the FM layer, which can be written as

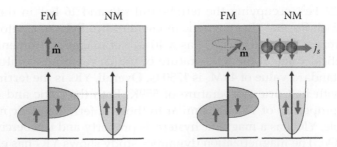

FIGURE 5.4
Spin pumping in a ferromagnetic (FM)/normal metal (NM) bilayer structure.

$$\alpha_{sp} = \frac{g\mu_B}{4\pi M_s} \frac{\text{Re}(g_{\uparrow\downarrow})}{d} \tag{5.2}$$

where μ_B is the Bohr magneton, g is the Landé factor, $4\pi M_s$ is the saturation induction, and d is the thickness of the FM layer.

5.1.2.2 Spin Hall and Inverse Spin Hall Effects

When an electron passes through the electric field of a nonmagnetic ion, it feels an effective magnetic field. Considering an electron and an ion in an NM thin film element as shown in Figure 5.5, this effective field can be written as[13]

$$\mathbf{B} = -\frac{1}{c^2} \mathbf{v} \times \mathbf{E} = -\frac{1}{c^2} \mathbf{v} \times \left(\frac{kq}{r^3}\mathbf{r}\right) \tag{5.3}$$

where $\mathbf{v}$ is the velocity of the electron, $\mathbf{E}$ is the electric field of the ion, and $\mathbf{r}$ is the position vector of the electron relative to the ion. Assuming the magnetic moment of the electron is μ, the alignment energy of the electron in field $\mathbf{B}$ can be written as[14]

$$H = -\mu \cdot \mathbf{B} = -(\mu_B \hat{\sigma}) \cdot \left[-\frac{1}{c^2} \mathbf{v} \times \left(\frac{kq}{r^3}\mathbf{r}\right)\right] = \frac{\mu_B kq}{c^2 r^3} (\mathbf{v} \times \mathbf{r}) \cdot \hat{\sigma} \tag{5.4}$$

where the unit vector denotes the direction of the electron's magnetic moment.

For the configuration in Figure 5.5, one has the vector $\mathbf{v} \times \mathbf{r}$ pointing up. For an electron with $\hat{\sigma}$ pointing up, the energy H is lower if the trajectory of the electron bends to the left. In contrast, for an electron with $\hat{\sigma}$ pointing down, the energy H is lower if the trajectory bends to the right. The net effect is a spin current in a direction transverse to the electron initial velocity and the accumulations of spin-up and spin-down electrons on the opposite edges of the film element. This effect is referred to as the spin Hall effect.[15–17] Inversely, a spin current can excite an electric current or voltage in a direction transverse to itself. This phenomenon is referred to as the inverse spin Hall effect.[18]

FIGURE 5.5
Spin Hall effect in a normal metal element.

It is important to note that the cartoons in Figure 5.5 correspond to a situation where the width of the NM element is on the same order as the spin diffusion length λ_{sd}, while the thickness of the NM element is significantly smaller than λ_{sd}. In the case where the width is significantly larger than λ_{sd} and the thickness is on the same order as λ_{sd}, one has only a spin current along the thickness direction and spin accumulations on the top and bottom surfaces.

5.1.2.3 Spin Seebeck Effect

Figure 5.6 illustrates the spin Seebeck effect[2,3,19] in a magnetic film strip. The film strip is magnetized and has a temperature gradient ΔT along its length direction. If ΔT is uniform along the x axis, the temperature of phonons $T_p(x)$ follows a linear profile. If the relaxation time in the magnon (or spin-wave) subsystem is significantly shorter than the magnon-phonon relaxation time, the temperature for magnons $T_m(x)$ can be considered constant along the x axis. In the case that the magnon-phonon relaxation is nontrivial and needs to be counted, the profile $T_m(x)$ is not constant and changes in a way so that the difference between $T_p(x)$ and $T_m(x)$ follows a hyperbolic sine function.[20] The difference $T_p(x) - T_m(x)$ gives rise to a spin voltage $\mu(x)$ in the film.[20,21] This voltage is positive at the cold end of the film strip and is negative at the hot end. If one places an NM thin film on the end of the FM film strip, the spin voltage can produce a spin current in the NM film[21]:

$$\langle J_s \rangle_z = \frac{g\mu_B \operatorname{Re}(g_{\uparrow\downarrow})k_B}{2\pi M_s d}(T_m - T_p) = 2\alpha_{sp}k_B(T_m - T_p) \tag{5.5}$$

This spin current can then produce an electric voltage across the NM film via the inverse spin Hall effect.

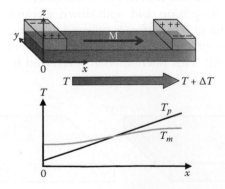

FIGURE 5.6
Spin Seebeck effect in a ferromagnetic film.

5.2 Spin Pumping

The effectiveness of spin angular momentum transfer across YIG/NM interfaces is probably the most fundamental issue to be addressed in the emerging research field of YIG-based spintronics. One typically uses the spin mixing conductance $g_{\uparrow\downarrow}$ to describe the interfacial momentum transfer efficiency, and the value of $g_{\uparrow\downarrow}$ depends critically on the spin transmission and reflection parameters at the interface. The spin mixing conductance has been determined experimentally for many FM/NM interfaces, and the experimental values have often agreed with theoretical predictions.[22-24] The spin mixing conductance at the YIG/NM interfaces might differ significantly from that at the FM/NM interfaces. This is because at the interface the transmission parameters are zero and the absolute values of the reflection parameters are one, due to the insulating feature of the YIG layer. Also, these conditions make the phases of the reflection parameters a very important factor. The transfer of spin angular momentum at YIG/NM interfaces is realized through *s-d* exchange interactions, where *s* refers to the conduction electrons in the NM layer and *d* refers to the local electrons in the YIG layer.[25]

5.2.1 Overview

As introduced in Section 5.1, the spin current density vector is expressed in terms of Equation (5.1). Theoretical study indicates the real part of spin mixing conductance plays the dominant role in angular momentum transfer at the FM/NM interface. $\text{Re}(g_{\uparrow\downarrow})$ is defined by the equation

$$\text{Re}(g_{\uparrow\downarrow}) = \frac{1}{2}\sum_n \left(\left| \mathbf{r}_n^\uparrow - \mathbf{r}_n^\downarrow \right|^2 + \left| \mathbf{t}_n^\uparrow - \mathbf{t}_n^\downarrow \right|^2 \right) \tag{5.6}$$

where $\mathbf{r}$ and $\mathbf{t}$ are the electrons' reflectivity and transmission coefficients at the FM/NM interface, respectively. The sum variable, n, refers to the number of transport channels, which was studied in the quantum electric transport theory. Simply speaking, if the transport is a ballistic type, which means no scattering region, there is only one left or right moving state on each energy level. This ideal situation is called single-channel model, that is, $n = 1$. Corresponding to the energy band graph, the electrons have only one path to move from the left/right chemical energy level to the right/left level.

However, in the real system, the interfacial electrons have multiple transverse modes of moving from the left/right lead to the right/left lead. Each transverse mode defines a transport channel. That means the electrons have n ($n > 1$) ways to move from one energy level to another in an energy band graph.

However, Equation (5.6) is only true when the FM layer is conductive. For the insulating FM layer, such as YIG, no electron can transmit into the FM layer, and the transmission coefficient t thereby equals zero. On the other hand, the reflectivity coefficient has the amplitude of 1 and is expressed in the form of $\mathbf{r}_n^{\uparrow\downarrow} = 1 \times e^{i\varphi^{\uparrow\downarrow}}$. Hence, the real part of spin mixing conductance in the insulating FM/NM structure can be simplified to

$$\mathrm{Re}(\mathbf{g}_{\uparrow\downarrow}) = \sum_n \left(1 - \cos(\varphi_n^\uparrow - \varphi_n^\downarrow)\right) \tag{5.7}$$

However, it is still hard to achieve $\mathrm{Re}(\mathbf{g}_{\uparrow\downarrow})$ by measuring the phase difference between spin-up and spin-down electrons. Alternately, let's look through the physics behind the spin pumping effect. The angular momentum transfer creates a torque, and such torque can influence the magnetization precession in the FM layer. The net effect is the damping enhancement. There must be some sort of connection between the efficiency of transferred angular momentum and the spin pumping-induced damping. Based on Equation (5.1), one can derive the expression of spin pumping-related damping. Spin current $\mathbf{J}_s$ is defined as the rate of pumped angular momentum per unit area. Then the rate of pumped angular momentum density per unit magnetization can be written as

$$\frac{\mathbf{J}_s}{M_s \cdot 1 \cdot d} = \frac{\hbar}{4\pi M_s} \mathrm{Re}(g_{\uparrow\downarrow}) \frac{1}{d} \hat{\mathbf{m}} \times \frac{\partial \hat{\mathbf{m}}}{\partial t} \tag{5.8}$$

where d and M_s are the thickness and saturation magnetization of the FM layer, respectively. Multiply $-|\gamma|$ on both sides of the angular momentum dynamic equation

$$\frac{\partial \hat{\mathbf{l}}}{\partial t} = \mathbf{m} \times \mathbf{H} - \frac{\hbar}{4\pi M_s} \mathrm{Re}(g_{\uparrow\downarrow}) \frac{1}{d} \hat{\mathbf{m}} \times \frac{\partial \hat{\mathbf{m}}}{\partial t} \tag{5.9}$$

One can get

$$\frac{\partial \mathbf{m}}{\partial t} = -|\gamma| \mathbf{m} \times \mathbf{H} + \frac{\hbar |\gamma|}{4\pi M_s} \mathrm{Re}(g_{\uparrow\downarrow}) \frac{1}{d} \hat{\mathbf{m}} \times \frac{\partial \hat{\mathbf{m}}}{\partial t} \tag{5.10}$$

Compare it with the torque equation:

$$\frac{\partial \mathbf{m}}{\partial t} = -|\gamma| \mathbf{m} \times \mathbf{H} + \alpha_{sp} \hat{\mathbf{m}} \times \frac{\partial \hat{\mathbf{m}}}{\partial t} \tag{5.11}$$

The spin pumping-induced damping can be written as

$$\alpha_{sp} = \frac{\hbar |\gamma|}{4\pi M_s} \mathrm{Re}(g_{\uparrow\downarrow}) \frac{1}{d} \quad \text{or} \quad \alpha_{sp} = \frac{g\mu_B}{4\pi M_s} \frac{\mathrm{Re}(g_{\uparrow\downarrow})}{d} \tag{5.12}$$

where μ_B is the Bohr magneton, and g, $4\pi M_s$, and d are the Landé factor, saturation induction, and thickness of the FM layer, respectively. The equation above gives the quantitative relationship between the momentum transfer and the spin pumping-induced damping constant of the FM layer. Finally, one can calculate the spin mixing conductance through the measurements of spin pumping-induced damping.

5.2.2 Spin Pumping in Nano-YIG/Au

Spin pumping is an interfacial effect, which means the strong effect can only be observed on the nanometer scale. The low-damping nano-YIG fabricated by Y. Sun et al.[26] has been used to study spin pumping at the magnetic insulator (MI)/NM interface.[27]

Nine-nanometer-nm thick YIG films were successfully deposited on $Gd_3Ga_5O_{12}$ (GGG) single-crystal substrate by a pulsed laser deposition (PLD) system, with optimized deposition conditions. B. Heinrich et al. built the multilayer structures on top of the bare YIG by the molecular beam epitaxy technique. The whole structure is GGG/YIG/Au/Fe/Au. The spin pumping effect is observed at the YIG/Au interfaces. The Fe layer acts as a spin sink in order to allow continuous generation of the spin current.

To determine the spin mixing conductance at the YIG/Au interface, one only needs to measure the spin pumping-induced damping. In this section, spin pumping is considered the only contribution to the additional damping. The spin pumping-induced damping is thereby simply equal to the damping of the whole YIG/Au/Fe/Au structure minus the bare YIG damping. The ferromagnetic resonance (FMR) technique is applied to measure the damping.

The FMR linewidth has a linear response to the microwave frequency, and the fitted slope is proportional to the damping, as shown in Equation (5.13):

$$\Delta H_{FMR} = \Delta H_0 + \frac{1}{\sqrt{3}} \frac{2\alpha}{|\gamma|} \frac{\omega}{2\pi} \qquad (5.13)$$

In Figure 5.7, the solid circle stands for the FMR data of bare YIG samples, while the solid square refers to the data from the YIG/Au/Fe/Au multilayer structure. Note that only the YIG FMR is discussed in this measurement. The ferromagnetic Fe layer has its own FMR resonance field and linewidth, but it doesn't affect spin pumping at the YIG/Au interface. After knowing the spin pumping-induced damping, the real part of spin mixing conductance can be calculated by Equation (5.12), where $d = 9$ nm, $g = 2.027$, $\mu_B = 9.274 \times 10^{-21}$ erg/G, and $4\pi M_s = 1.31$ kG. The calculated spin mixing conductance for YIG/Au is $1.2 \times 10^{14}/cm^2$. This value is about one order smaller in comparison with the spin mixing conductance of the metallic FM/NM interface. The much smaller spin mixing conductance in MI/NM

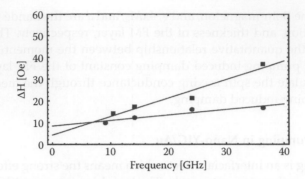

FIGURE 5.7
Frequency-dependent FMR linewidth for bare YIG and YIG/Au. (Reprinted with permission from B. Heinrich, et al., *Phys. Rev. Lett.* 107, 066604 (2011). Copyright 2011 by the American Physical Society.)

is due to the zero electrons transmission at its interface, which reduces the efficiency of angular momentum transfer.

The spin pumping in YIG/Au can be enhanced by chemical modifications of the YIG surface using Ar^+ ion beam etching.[28] After low-energy and low-incidence angle etching at 400°C, the spin mixing conductance increases by a factor of five in comparison to the untreated YIG/Au interface. The enhancement of spin pumping is related to the change in chemical states of Fe and O, which were measured by x-ray photoelectronic spectrometry.

The results in Table 5.1 indicate that spin pumping-induced damping increased significantly using Ar^+ etching.

5.2.3 Spin Pumping in Nano-YIG/Pt

In the previous subsection, the Fe layer acts as the spin sink to guarantee the continuous flow of spin current in the YIG/Au/Fe/Au multilayer. The ferromagnetic Fe exchanged coupled with YIG, and therefore, makes the

TABLE 5.1

Magnetic Parameters for the Measured Samples before and after Etching

Sample	$g^{\uparrow\downarrow}$ $10^{14}\,cm^{-2}$	$4\pi M_{eff}$ kOe	$\alpha_{untreated}$ 10^{-3}	α_{etched} 10^{-3}	α_s 10^{-3}	α_d 10^{-3}
S1		1.958	0.5			
S2	0.9	1.706			0.7	
S3		1.728	0.5	0.3	1.4	
S4	4.9	2.116	0.5	0.5		6.9
S5	5.0	1.997		0.3		6.9
S6	3.6	2.031	0.3	1.4		6.1

Source: Reprinted with permission from C. Burrowes, et al., *Appl. Phys. Lett.* 100, 092403 (2012). Copyright 2012 by AIP Publishing LLC.

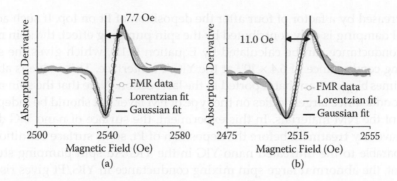

FIGURE 5.8
FMR profile for (a) bare YIG(25 nm) sample and (b) YIG(25 nm)/Pt(20 nm) sample. (Reprinted with permission from Y. Sun, et al., *Phys. Rev. Lett.* 111, 106601, 2013. Copyright 2013 by the American Physical Society.)

situation complex. One of the influences is the shift in the YIG resonance field. To avoid such a ferromagnetic spin sink, some nonmagnetic heavy metals can be used. Platinum (Pt) is one of the ideal candidates. In addition, Pt has already drawn attention in the spin Hall effect study due to its strong spin-orbit interaction. The spin pumping effect in the nano-YIG/Pt bilayer structure will be reviewed in this subsection and Section 5.3.

The spin current is generated at the YIG/Pt interface and furthermore enhances the magnetic damping. Figure 5.8 shows the FMR profiles for both bare YIG and YIG/Pt samples, which were measured in the X band cavity. The experimental data were fitted by derivative Lorentzian and Gaussian functions. Both fittings indicate that the FMR linewidth of YIG/Pt is much bigger than that of bare YIG.

The damping results are shown in the Figure 5.9. The FMR measurements were carried out in the Ku band shorted waveguide, and the frequency range is from 13 to 18 GHz. One can see that the YIG damping is increased significantly after the capping of the Pt layer. The dampings of YIG and YIG/Pt are 8.7×10^{-4} and 35.1×10^{-4}, respectively. This indicates the YIG's damping

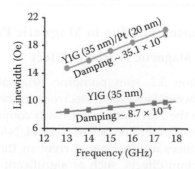

FIGURE 5.9
Damping for (a) bare YIG(35 nm) sample and (b) YIG(35 nm)/Pt(20 nm) sample.

is increased by a factor of four after the deposition of Pt on top. If such additional damping is only contributed by the spin pumping effect, the spin mixing conductance can be calculated by Equation (5.12), which gives the spin mixing conductance of 6.4×10^{14} at the YIG/Pt interface. This value is about five times bigger than that reported in the last section. Note that the spin mixing conductance largely relies on the type of FM layer but should be independent of the NM materials. In this experiment, the surface of nano YIG does not have any treatment before the deposition of Pt, so its surface condition is comparable to the untreated nano YIG in the YIG/Au spin pumping study. In that, the abnormal large spin mixing conductance in YIG/Pt gives rise to the hypothesis of the existence of an additional damping different from spin pumping-induced damping. The new damping is believed to have originated from the dynamic exchange coupling at YIG/Pt interfaces caused by the magnetic proximity effect. Section 5.3 focuses on this new damping in detail.

In this section, we study the spin pumping effect at the nano-YIG/NM interface. The important physics parameter, spin mixing conductance, was determined by FMR measurements. The value of spin mixing conductance in nano-YIG/NM is only 1/10 that in Fe/NM. To enhance the efficiency of angular momentum transfer at the interface, one can use the low-energy and low-incident-angle Ar+ ion beam to etch the nano-YIG surface. This etching changes the chemical state of Fe and O. Furthermore, it leads to the enhancement of the spin pumping effect at the nano-YIG/Au interface. Heavy metal Pt is another NM to be used in the spin pumping study. We observed a significant increase in damping in the YIG/Pt bilayer structure. The damping of YIG/Pt might have four contributions: (1) the Gilbert or intrinsic damping of nano-YIG, (2) surface defects-associated two-magnon scattering, (3) spin pumping-induced damping, and (4) a new damping, which will be discussed in the next section.

5.3 Damping Enhancement due to Magnetic Proximity Effect

5.3.1 Introduction to Magnetic Proximity Effect

As discussed in Section 5.2, spin pumping originates from dynamical exchange coupling between localized d electrons in the FM component and itinerant s electrons in the NM component. Such coupling gives rise to the transfer of spin angular momentum across the FM/NM interface. This spin transfer not only produces a pure spin current in the NM layer, but also results in other important effects, such as significant enhancement in the damping of the FM layer. The strong interest in the spin pumping effect not only is driven by the intriguing fundamental physics associated with the effect, but also is due to the promising future applications of the effect in spintronic devices, such as spin batteries.

Many FM/NM systems also show a magnetic proximity effect where magnetic moments are induced in several NM atomic layers that are in close proximity to the FM component.[29-36] The magnetic proximity effect (MPE) originates from direct static exchange coupling between the NM itinerant electrons and the FM moments at the interface. The induced moments consist of contributions from both the spin and orbital motions of the electrons. The MPE has been experimentally observed in a number of FM/NM systems, including Fe/Pd,[29-31] $Ni_{81}Fe_{19}$/Pd,Ni/Pt,[32] $Ni_{80}Fe_{20}$/Pt,$Y_3Fe_5O_{12}$(YIG)/Pt,[36] Fe/W,[33] Fe/Ir,[33] and Co/Au.[34] In some of these systems, the induced moment per atom has also been determined by x-ray magnetic circular dichroism (XMCD) measurements. In particular, 0.17–0.29 μ_B/atom has been reported for Pt atoms in proximity to Ni,[32] 0.2 μ_B/atom has been measured for W and Ir atoms in proximity to Fe,[33] 0.031 μ_B/atom has been reported for Au atoms in proximity to Co,[34] and very recently, 0.054 μ_B/atom has been determined for Pt atoms in proximity to YIG.[36]

Although the above-mentioned moments are significantly smaller than the moments per atom for ferromagnetic elements in ferromagnets, they play critical roles in many important effects. For example, previous experiments have demonstrated that the MPE was responsible for enhanced magnetic moments in Fe/Pd, $Ni_{81}Fe_{19}$/Pd, and Ni/Pt superlattices, ferromagnetism in submonolayer Fe films grown on Pd substrates,[30,31] the anomalous Nernst effect in YIG/Pt structures, ferromagnetic coupling in $Ni_{80}Fe_{20}$/Pt/$Ni_{80}Fe_{20}$ that differs from the RKKY coupling, and the modification of the magneto-optic spectra of Co in Co/Au multilayers[37] and of Fe in Fe/Au multilayers.[38]

5.3.2 MPE Caused Additional Damping in YIG/Pt

Recently, Y. Sun et al. reported the experimental results of enhanced damping in nanometer-thick YIG films capped by Pt,[39] which is believed to have originated from the ferromagnetic ordering in Pt atomic layers in close proximity to the YIG/Pt interface and the dynamic exchange coupling between the ordered Pt spins and the spins in the YIG film. Three points should be emphasized. (1) The magnetic order of the Pt layer is caused by the magnetic proximity effect, which has been confirmed by XMCD measurement.[36] (2) The dynamic exchange coupling at the YIG/Pt interface results in the transfer of partial damping of Pt to YIG, and thereby the damping enhancement. (3) Due to the presence of ferromagnetic ordering of the Pt layer, traditional spin pumping at the FM/NM interface does not exist. Instead, the spin pumping occurs at the interface between FM Pt and nonmagnetic Pt, which contributes to the damping of FM Pt.

In the experiment, YIG nanometer-thick films were capped by Pt layers. When the Pt was thicker than 3 nm, a magnetic proximity effect-induced damping α_{MPE} was observed. This new damping is significantly larger than both the bare YIG damping and the expected spin pumping damping. After inserting a Cu spacer between YIG and Pt, the new damping α_{MPE} can be suppressed. Four groups of samples were fabricated for this study: (1) YIG/Pt with Pt thickness in the range of 0 ~ 20 nm, (2) YIG/Cu as the control samples, (3) YIG/Cu/Pt with the thickness variation of a Cu spacer, and (4) YIG/Cu*/Pt, where Cu* means the natural oxidized Cu spacer. Specially, the samples in groups 1–3 were deposited by PLD, while the YIG and Pt layers of samples in group 4 were grown by PLD, but their Cu spacers were deposited by an AJA magnetron sputtering system.

Figure 5.10 shows the damping change with Pt thickness in the YIG/Pt samples. FMR measurements were carried out in the Ku band shorted waveguide. Figure 5.10(a) shows the frequency-dependent peak-to-peak linewidth. The effective damping is proportional to the slope of the linear fit. Figure 5.10(b) shows the damping as a function of Pt thickness. The bare YIG damping is as low as 3.6×10^{-4}, which gives the baseline of the damping comparison shown in Figure 5.10(b). For Pt thinner than 3 nm, the damping of YIG/Pt increases dramatically. There exists a damping saturation for samples with thicker Pt. As one increases the thickness of the Pt layer (from Pt clusters to a more or less continuous film and then to a better epitaxial film), the proximity effect develops gradually and becomes stronger. As a result, the dynamic coupling-produced damping increases with the Pt thickness (d_{Pt}). This explains the responses shown in Figure 5.10(b) for $d_{Pt} \leq 3$ nm. For $d_{Pt} > 3$ nm, however, an increase in d_{Pt} leads to no further changes because the ordering of electrons takes place only in the first several atomic layers of Pt, and the growth of additional Pt atomic layers has no effect on the ordering in the first several layers. This explains the saturation responses shown in Figure 5.10(b) for $d_{Pt} > 3$ nm. Overall, there are two contributions to the effective damping: (1) bare YIG damping α_0 and (2) damping due to magnetic proximity effect.

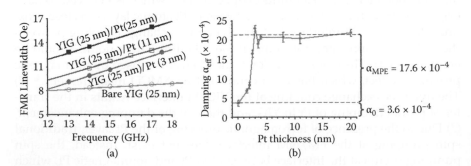

FIGURE 5.10

(a) ΔH versus ω responses for bare YIG(25 nm) and YIG(25 nm)/Pt(t nm), t = 3, 11, 20. (b) Damping α_{eff} as a function of the Pt thickness. (Reprinted with permission from Y. Sun, et al., *Phys. Rev. Lett.* (in press). Copyright 2013 by the American Physical Society.)

To further test the interpretation, the authors inserted a Cu spacer between YIG and Pt to decouple the interlayer dynamic exchange coupling and thereby suppress the new damping α_{MPE}. Two kinds of Cu spacers were applied: (1) pure Cu and (2) Cu with natural oxidized surface. Both spacers can separate the Pt layer from YIG thin films. In that, the α_{MPE} will disappear due to the absence of the magnetic proximity effect. For the pure Cu spacer, spin pumping occurs at the YIG/Cu interface and the spin pumping-induced damping α_{SP} will rise. For the surface-oxidized Cu spacer, however, the oxidation results in high resistance in the spacer, and hence causes blocking of the spin current flow. One can expect (1) the suppression of α_{MPE} but detectable α_{SP} in the samples with pure Cu spacer, and (2) the complete suppression of both α_{MPE} and α_{SP} with surface-oxidized Cu spacer. The authors indeed confirmed these two expectations in the following experiments.

The first expectation is confirmed by the damping measurements on YIG (35 nm)/Cu/Pt(23 nm) and YIG(35 nm)/Cu samples, where the latter are the controls. The thickness of Cu varies in the range of 0 ~ 18 nm. Figure 5.11 shows the effective damping change with the Cu thickness, d_{Cu}. For the YIG/Cu/Pt samples, the damping drops significantly with the increase in Cu thickness when $d_{Cu} < 10$ nm. Then the reduction shows saturation. For the control samples YIG/Cu, the damping variation is really tiny in the d_{Cu} range of 0 ~ 13 nm. The authors give the interpretation of the data: (1) The effective damping reduction is caused by the decoupling of a dynamic exchange interaction between FM ordering Pt and YIG through the Cu spacer. (2) When the Cu is thick enough, the α_{MPE} will be completely suppressed and the reduction is saturated. (3) The damping gap between YIG/Cu/Pt and YIG/Cu is attributed to spin pumping. (4) The effective damping of the YIG/Cu control is slightly higher than the bare YIG α_0 due to the interface conditions, such as interfacial roughness. One can see that the spin pumping only occurs in the YIG/Cu/Pt samples, and not the YIG/Cu control. The reason is that several nanometer-thick Cu cannot act as a good

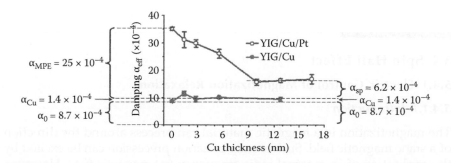

FIGURE 5.11
Effective damping as a function of Cu thickness. (Reprinted with permission from Y. Sun, et al., *Phys. Rev. Lett.* 111, 106601, 2013. Copyright 2013 by the American Physical Society.)

TABLE 5.2

Effective Damping Parameters

Sample	α_{eff} (×10⁻⁴) (in-plane)	α_{eff} (×10⁻⁴) (out-of-plane)
YIG(19 nm)	3.58 ± 0.76	4.72
YIG(19 nm)/Pt(20 nm)	35.36 ± 0.58	36.60
YIG(19 nm)/Cu*(5 nm)/Pt(20 nm)	4.33 ± 0.58	3.41
YIG(19 nm)/Cu*(5 nm)	4.82 ± 0.62	6.71

Source: Reprinted with permission from Y. Sun, et al., *Phys. Rev. Lett.* 111, 106601, 2013. Copyright 2013 by the American Physical Society.

spin sink due to its large spin diffusion length (>500 nm). For the YIG/Cu/Pt samples, however, the spin sink was provided by the Pt layer.

The second expectation was simply tested by inserting a 5-nm thick oxidized Cu spacer into YIG/Pt. Table 5.2 shows the effective damping parameters of bare YIG, YIG/Pt, YIG/Cu*/Pt, and YIG/Cu, where the Cu* refers to the oxidized Cu spacer. The thicknesses of YIG and Pt are 19 and 20 nm, respectively. One can see that effective damping is about 10 times larger than bare YIG damping. However, after the insertion of Cu*, it decreased to the same level as the YIG/Cu control. This result indicates the Cu* can completely suppress the α_{MPE} by decoupling the interfacial dynamic exchange interaction caused by the magnetic proximity effect. The spin pumping-induced damping α_{SP} is also switched off because the spin current is blocked by the Cu oxide spacer.

The confirmation of the two expectations above therefore supports the interpretation on the origin of α_{MPE}. The new damping we observed should be a common phenomenon in any FM/NM system where the magnetic proximity effect exists, including Ni/Pt, NiFe/Pt, Fe/Pd, NiFe/Pd, Fe/W, Fe/Ir, and Co/Au. Therefore, this work reveals a new channel for magnetic damping and will stimulate new theoretical studies on damping in FM/NM systems.

5.4 Spin Hall Effect

5.4.1 Electric Control of Magnetization Relaxation[*]

5.4.1.1 Introduction

The magnetization in a magnetic material can precess around the direction of a static magnetic field. Such a magnetization precession can be excited by the application of an external radio frequency (rf) magnetic field. However,

---

[*] Reprinted with permission from Z. Wang, et al., *Appl. Phys. Lett.* 99, 162511 (2011). Copyright 2011 by AIP Publishing LLC.

once the rf field is removed, the magnetization will relax back to its equilibrium direction. This magnetization relaxation can be due to energy redistribution within the magnetic subsystem, energy transfer out of the magnetic subsystem to nonmagnetic subsystems such as phonons and electrons, or energy transfer out of the material to external systems.[12,40]

One can change the rate of the magnetization relaxation of a magnetic material by controlling the material fabrication processes. However, once the material is made, the relaxation rate is generally considered unchangeable at a certain external field. In contrast, recently two approaches have been demonstrated that can control magnetization relaxations in magnetic thin films: one makes use of the flow of spin-polarized electrons through the films,[41,42] and the other takes advantage of the injection of spin-polarized electrons into the films.[43,44] Although both approaches rely on angular momentum transfer between the polarized conduction electrons and the spins in the films, there exists a substantial difference: the first approach involves the flows of net charge currents through the film materials, while the second does not. A demonstration of such relaxation control is of great significance, both fundamentally and practically. In practical terms, the control of magnetization relaxation is highly desirable because the magnetization relaxation not only plays a critical role in the dynamics of spin-based devices, but also sets a natural limit on the response time of a device and determines the magnetization noise.

The two approaches demonstrated so far apply to metallic films only. This section presents the control of magnetization relaxations in thin-film magnetic insulators. Specifically, it is demonstrated that one can control FMR linewidth in thin-film magnetic insulators via interfacial spin scattering (ISS). The experiments use nanometer-thick ferrimagnetic YIG films capped with a nanometer-thick Pt layer. An electric voltage signal is applied to the Pt layer that produces a spin current along the Pt thickness direction via the spin Hall effect.[15,16] As the spin current scatters off the surface of the YIG film, it exerts a torque on the YIG surface spins. Due to the exchange interactions, the effect of this torque is extended to other spins across the YIG thickness, and thereby changes the FMR response in the whole YIG film.

5.4.1.2 Experimental Configuration and Parameters

The net effect of the ISS process on the FMR response depends critically on the relative orientation between (1) the magnetic moments of the electrons in the Pt layer, which move toward the YIG surface; and (2) the precession axis of the magnetic moments in the YIG film. When they are antiparallel, the FMR linewidth is reduced, which indicates a decrease in the relaxation rate. In contrast, for a parallel configuration, the FMR linewidth is broadened, which indicates an increase in the relaxation rate. Moreover, the FMR linewidth versus spin current response shows a current threshold and nonlinear behavior. It is important to emphasize that as the parallel/antiparallel

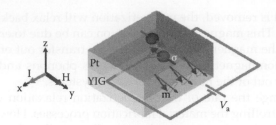

FIGURE 5.12
Experimental configuration for control of ferromagnetic resonance through interfacial spin scattering. (Reprinted with permission from Z. Wang, et al., *Appl. Phys. Lett.* 99, 162511 (2011). Copyright 2011 by AIP Publishing LLC.)

configuration can be changed simply by reversing the direction of the current or the field, the results demonstrate a simple approach for relaxation control in magnetic insulators.

Figure 5.12 shows a schematic of the experimental configuration. The YIG film is magnetized by an in-plane static magnetic field H. The Pt film is biased by a dc voltage V_a. This voltage signal leads to a current I along the $+x$ direction and a flow of electrons along the $-x$ direction, and the latter produces a spin current in the Pt thickness direction via the spin Hall effect. The spins moving toward the YIG film have their magnetic moments in the $-y$ direction, while those moving toward the Pt top surface have their moments in the $+y$ direction. In Figure 5.12, these directions are indicated by the short arrows in the Pt layer. The YIG/Pt element is placed inside a shorted rectangular waveguide, and the latter produces a microwave magnetic field h, which is in the film plane and perpendicular to H. For the discussions below, $+I$ denotes a dc current along the $+x$ direction, while $-I$ denotes a dc current along the $-x$ direction.

The YIG films were deposited with pulsed laser deposition (PLD) techniques. The deposition used a 1 in. diameter YIG target and a single-crystal (111) gadolinium gallium garnet (GGG) substrate. The target-to-substrate distance was kept constant at 7 cm. The deposition was done in high-purity (99.999%) oxygen for several minutes. Prior to the deposition, the system had a base pressure of 3.4×10^{-7} Torr. During the deposition, the substrate temperature was kept constant at 790°C and the oxygen pressure was 0.1 Torr. Right after the deposition, the film sample was annealed at the same temperature in the same oxygen atmosphere for 10 min. The cooling of the films was in a 400 Torr oxygen environment at a rate of 2°C/min. The deposition used 248 nm KrF excimer laser pulses with an energy fluence of 1.7 J/cm², a pulse duration of 30 ns, and a pulse repetition rate of 2 Hz. These laser parameters together with the target-to-substrate distance yielded a YIG growth rate of about 1 nm/min. After the deposition, each YIG film was capped by a Pt layer with the PLD technique. The Pt deposition was done at room temperature with a base pressure of 4.0×10^{-7} Torr and the other PLD parameters described above.

For the data presented below, the YIG/Pt element was 4.7 mm long and 3.7 mm wide, and the thicknesses of the YIG and Pt films were 10.0 and 5.6 nm, respectively, as determined by x-ray reflectivity measurements. Atomic force microscopy measurements indicated that the YIG film had a surface roughness of about 0.5 nm. Static magnetic measurements yielded a saturation induction of 1858 G. This value is about 6% larger than that for YIG bulk materials. Possible reasons for this difference include the error in film thickness determination and small deviations in chemical composition near YIG film surfaces. All FMR measurements were done by sweeping the field at a fixed microwave frequency of 11.5 GHz. The measurements used field modulation and lock-in detection techniques.

5.4.1.3 Experimental Results and Discussion

Figure 5.13 shows representative FMR profiles measured in the absence of a dc current ($I = 0$). Figure 5.13(a) shows the power absorption derivative profile. The profile in Figure 5.13(b) is obtained through the integration of the profile in Figure 5.13(a) with the field. The full linewidth at half maximum of the profile in Figure 5.13(b) is the so-called FMR linewidth ΔH. One can see that the ΔH value of the YIG film is much larger than that of YIG single crystals. Frequency-dependent FMR measurements indicated that this large value is mainly due to inhomogeneity-caused line broadening and two-magnon scattering.In the presence of a dc current I, the FMR linewidth is

$$\Delta H_I = \Delta H_0 + \Delta H_{heating} + \frac{2\omega}{\gamma}\Delta\alpha_I \qquad (5.14)$$

where ΔH_0 is the FMR linewidth measured for $I = 0$, $\Delta H_{heating}$ denotes the change in linewidth due to dc current-produced heating, and the last term describes the ISS-produced change in linewidth. In the last term,

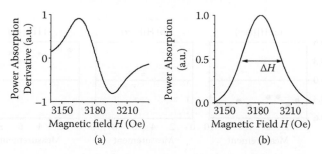

(a) (b)

FIGURE 5.13
Ferromagnetic resonance profiles: (a) derivative signal and (b) integrated intensity. (Reprinted with permission from Z. Wang, et al., *Appl. Phys. Lett.* 99, 162511 (2011). Copyright 2011 by AIP Publishing LLC.)

$\gamma = 2.8$ MHz/Oe is the absolute gyromagnetic ratio, and $\Delta\alpha_I$ denotes the ISS-caused change in Gilbert constant α. Note that the Gilbert constant has been widely used to describe magnetization relaxations,[40] and Equation (5.14) has assumed a Gilbert-like damping for the ISS contribution to the relaxation. It is important to emphasize that the value of $\Delta H_{heating}$ depends on the magnitude of the dc current, and not the direction of the current, while the value of $\Delta\alpha_I$ depends on both. One can in fact expect $\Delta\alpha_{+I} = -\Delta\alpha_{-I}$, as discussed below. For this reason, one can simply use the following equation to describe the ISS-associated damping:

$$\Delta H_I - \Delta H_{-I} = \frac{2\omega}{\gamma}(\Delta\alpha_{+I} - \Delta\alpha_{-I}) = \frac{4\omega\Delta\alpha_{+I}}{\gamma} \tag{5.15}$$

Figure 5.14(a)–(c) show $\Delta H_I - \Delta H_{-I}$ for field H along the $+y$, $+x$, and $-y$ directions, respectively, with each configuration measured ten times. For all measurements, the magnitude of the dc current was 80 mA. One can see in Figure 5.14 completely different values for different field configurations. All values in Figure 5.14(a) are negative and indicate a negative $\Delta\alpha_{+I}$ for the $H||(+y)$ configuration. In stark contrast, all values in Figure 5.14(b) for $H||(+x)$ are small and indicate $\Delta\alpha_{+I} \approx 0$, and all values in Figure 5.14(c) for $H||(-y)$ are positive and indicate $\Delta\alpha_{+I} > 0$.

These results can be interpreted as follows. When spin-polarized electrons scatter off the YIG surface, they transfer a certain angular momentum to the surface spins in the YIG film.[25] This momentum transfer is realized through the s-d exchange interactions at the Pt/YIG interface. Here, s refers to spin-polarized electrons in the Pt layer and d refers to localized electrons on the surface of the YIG film. The interfacial momentum transfer results in a net torque on the YIG surface spins[43,44]:

$$\tau = C\hat{m} \times \hat{\sigma} \times \hat{m} \tag{5.16}$$

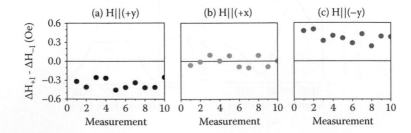

FIGURE 5.14
$\Delta H_{+I} - \Delta H_{-I}$ values measured at different times for different field configurations, as indicated. (Reprinted with permission from Z. Wang, et al., _Appl. Phys. Lett._ 99, 162511 (2011). Copyright 2011 by AIP Publishing LLC.)

where $\hat{m}$ denotes the magnetic moment direction of the YIG surface spins, $\hat{\sigma}$ denotes the magnetic moment direction of the spin-polarized electrons in the Pt film, which moves toward the YIG surface, and the coefficient C is associated with the strength of the spin current and the properties of the YIG and Pt layers. For the $H||(+y)$ configuration, the precession of $\hat{m}$ is around the $+y$ direction and the torque counters the damping of the $\hat{m}$ precession. The net effect is a decrease in damping. For the $H||(-y)$ configuration, in contrast, the precession of $\hat{m}$ is around the $-y$ direction and the torque leads to an additional damping. One can expect that the changes in the two cases have the same magnitude, even though they have opposite signs. This is because the magnitude is the same for both configurations. Thus, one can expect $\Delta\alpha_{+I} = -\Delta\alpha_{-I}$, as mentioned above. In the case of $H||(+x)$, the precession axis is perpendicular to $\hat{\sigma}$, and the average of the torque τ over each precession period is very small. As a result, the torque produces negligible effects. It is important to emphasize that although the torque τ is exerted on the surface spins in the YIG film, its effect is extended to other spins across the YIG film thickness due to exchange interactions. This is possible because the YIG film thickness is smaller than the exchange length. Indeed, a similar behavior has even been demonstrated in YIG films with a thickness of several microns.[1]

The $\Delta H_{+I}-\Delta H_{-I}$ values measured at different dc currents are shown in Figure 5.15(a). Each point shows the averaged value over five to ten measurements, and the error bar for each point shows the corresponding standard deviation. The right axis shows the corresponding $\Delta\alpha$ values evaluated by Equation (5.15). The top axis shows the corresponding spin current densities obtained by $J_s = \theta_{SH}J_c$, where $\theta_{SH} = 0.013$ was the Pt spin Hall angle and J_c was the charge current density. Three important results are evident in Figure 5.15(a). (1) There is a current threshold for the onset of the ISS effect. (2) Above the threshold, the sign of $\Delta\alpha$ agrees with the expectation discussed above. (3) Above the threshold, the magnitude of $\Delta\alpha$ increases nonlinearly with the spin current, with a negative curvature. Similar ISS effects were observed for other YIG/Pt samples. Figure 5.15(b) shows data for a YIG/Cu control sample. Each point shows the averaged value over five measurements, and the error bar for each point shows the corresponding standard deviation. Note that the thickness of the Cu layer is very close to that of the Pt layer in the YIG/Pt sample. Because the spin Hall effect in Cu is very weak,[46] the sample showed no ISS effects. This indicates that the spin current in the Pt layer is critical for the ISS effects presented.

The Oersted field produced by the dc current in the Pt layer was estimated to be less than 0.2 Oe. This field was significantly smaller than the FMR fields, which were about 3,200 Oe. Although the Oersted field did cause a slight shift in the FMR field, it resulted in negligible effects in ΔH because the FMR frequency was kept constant. The observed threshold effects are rather unexpected. One possible origin is the change of spin accumulations in the

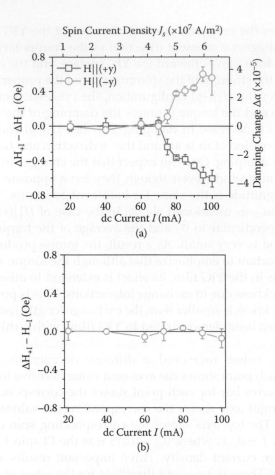

FIGURE 5.15

(a) ΔH_{+I}–ΔH_{-I} as a function of dc current I and spin current density J_s for the YIG/Pt sample. The right axis shows the corresponding change in damping constant α. (b) ΔH_{+I}–ΔH_{-I} as a function of dc current I for a YIG/Cu control sample. The sample is 5.0 mm long and 3.7 mm wide. The YIG film is 6 nm thick, and the Cu capping layer is 5 nm thick. The magnetic field is along the +y direction. (Reprinted with permission from Z. Wang, et al., *Appl. Phys. Lett.* 99, 162511 (2011). Copyright 2011 by AIP Publishing LLC.)

Pt layer with temperature. At low dc currents, the spin diffusion length, which is about 10 nm,[47] is relatively large in comparison with the Pt thickness, and the efficiency in building spin accumulations near the Pt surfaces is relatively small. As the dc current is increased to a certain level, the current-induced heating leads to a decrease in the spin diffusion length[15] and a corresponding increase in the efficiency of building the spin accumulations.

In summary, this subsection presented the electric control of magnetization relaxation in YIG thin films via the ISS process. It was found that the ISS effect can play a positive or negative role in the relaxation, and one can control this role by simply changing the strength and direction of the dc current. It was also found that the damping constant versus spin current response showed a current threshold and nonlinear behavior. Future work that is of great interest includes the study of the roles of YIG and Pt thicknesses on the ISS effects and the demonstration of ISS effects in other materials. Future study on the origins and features of the observed threshold and nonlinear effects is also of great interest.

5.4.2 Control of Spin-Wave Amplitude in Ferrimagnetic Insulator

5.4.2.1 Spin-Wave Damping and Parametric Pumping

Spin waves in ferromagnetic films have many unique properties, and thereby have potential for applications in microwave signal processing,[8-10] logic operations,[48-50] and insulator-based electrical signal transmissions.[1] These applications, however, are bottlenecked by the damping of spin waves. Such damping can result from various physical processes, such as spin-orbit coupling, scattering on defects, and three- and four-wave nonlinear interactions.

One way to compensate for spin-wave damping is to use parametric pumping.[51] Previous experiments have demonstrated that the spin waves in magnetic thin films could be parametrically amplified.[52,53] The spin-wave traveling in the YIG film loses energy during propagation. The energy lost is compensated by an external microwave source through three-wave parametric pumping. This method, however, requires the use of (1) an external microwave signal with a frequency twice that of the spin wave and (2) a delicate microwave resonator structure for the delivery of this signal to the magnetic film. Moreover, the amplification is limited to a very narrow frequency range, which is determined by the frequency conditions of the parametric resonance.

5.4.2.2 Amplification of Spin Waves through ISS

The work described in Chapter 4 demonstrates the control of FMR linewidth through the interfacial spin scattering effect. This chapter presents the results on the amplification of spin waves. Experiments use a 4.6-μm-thick YIG film strip with a 20-nm-thick Pt capping layer. A dc current pulse is applied to the Pt film and produces a spin current along the Pt thickness direction via the spin Hall effect.[15-17] As the spin current scatters off the surface of the YIG film, it exerts a torque on the YIG surface spins. Due to the dipolar and exchange interactions, the effect of this torque is extended to other spins across the YIG thickness, and thereby to spin-wave pulses that travel in the YIG film.

The net effect of the ISS process on spin waves depends critically on the relative orientation of (1) the magnetic moments of the electrons in the Pt layer, which scatter off the YIG surface, and (2) the precession axis of the magnetic moments on the YIG surface. When they are antiparallel, the spin-wave damping is reduced and the amplitude of a traveling spin-wave pulse is increased. In contrast, in a parallel configuration, the spin-wave pulse experiences an enhanced attenuation. The ISS process can also raise or reduce the power level to which high-power spin-wave pulses saturate due to nonlinear damping.[51,54]

It is important to emphasize that as the parallel/antiparallel configuration can be changed simply by reversing the direction of the dc current, this work demonstrates a rather simple new approach for the control of spin waves. One can expect that in the future, this ISS effect would allow for the realization of decay-free spin-wave propagation and the development of a new class of electronic devices.

5.4.2.3 Experimental Configuration and Parameters

Figure 5.16 shows a schematic of the experimental configuration. The core component is a long and narrow YIG film strip with its central portion covered by a Pt thin film. The YIG film is magnetized by an in-plane magnetic field H. The field is referred as a positive field ($H > 0$) when it is applied along the $+y$ direction and a negative field ($H < 0$) when it is in the $-y$ direction. This film/field configuration supports the propagation of *surface* spin waves, of which the amplitude has an exponential distribution along the YIG film thickness. When $H > 0$, the spin wave with a wave vector **k** along the $+x$ direction has a larger amplitude near the top surface of the YIG film. When $H < 0$, the spin wave along the $+x$ direction has a larger amplitude near the bottom surface of the YIG film. Two microstrip transducers are placed on the right and left ends of the YIG strip for the excitation and detection, respectively, of the spin wave.

A dc voltage V_a is applied to the Pt film. A positive voltage ($V_a > 0$) results in a current flow along the $+x$ direction and a flow of electrons along the $-x$ direction. The electrical current produces a spin current along the Pt thickness direction via the spin Hall effect. When $V_a > 0$, the electrons moving toward the YIG film have their magnetic moments in the $-y$ direction. In Figure 5.16, the small spheres in the Pt layer represent electrons, and the

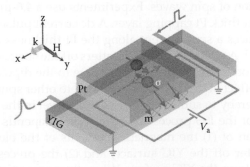

FIGURE 5.16

Experimental setup for control of spin waves through interfacial spin scattering in a YIG/Pt structure. (Reprinted with permission from Z. Wang, et al., *Phys. Rev. Lett.* 107, 146602 (2011). Copyright 2011 by the American Physical Society.)

arrows through the spheres indicate the directions ($\hat{\sigma}$) of the electron magnetic moments.

For the data presented below, the YIG strip was 4.6 μm thick, 2.2 mm wide, and 22 mm long. It was cut from a larger single-crystal (111) YIG film grown on a gadolinium gallium garnet substrate by liquid phase epitaxy. The Pt film was grown on the YIG strip at room temperature by PLD. Based on the PLD parameters, the thickness of the Pt film was estimated to be about 20 nm. This value matches the estimation based on the resistance of the Pt element, which was 20.1 nm. For this estimation, a resistivity of $\rho = 371$ nΩ·m was used for the Pt film. The Pt element had the same width as the YIG strip and a length of $L = 3.5$ mm. The microstrip transducers were 50 μm wide and 2.0 mm long. The transducer separation was held at 5.5 mm, with each transducer 1.0 mm away from the Pt element.

The signals applied to the excitation transducer were microwave pulses with a width of 50 ns and a repetition period of 10 ms. These signals excited spin-wave pulses in the YIG strip. The signals applied to the Pt element were dc pulses with a width of 300 ns and the same period as the microwave pulses. The microwave pulses had a delay of 20 ns relative to the dc pulses. These parameters ensure that the spin current was on over the entire propagation time of each spin-wave pulse.

5.4.2.4 Experimental Results and Discussion

Figure 5.17 presents the data measured for $H = 683$ Oe. Graph (a) shows the transmission profile for the transducer-YIG-transducer structure. The dashed line indicates a frequency of 3.636 GHz, which was the carrier frequency of the input microwave pulses for most of the measurements. Graph (b) shows output signals measured for three different dc pulse voltages (V_a) applied to the Pt layer. The time $t = 0$ corresponds to the moment when a microwave pulse enters the excitation transducer. Graph (c) gives the relative change in the peak voltage of the output signal as a function of V_a. Here, the relative change is defined as $(V - V_0)/V_0$, where V_0 is the peak voltage of the output signal for $V_a = 0$ and V is the peak voltage of the output signal for $V_a \neq 0$. For the data in both (b) and (c), the power of the input microwave pulses was set to 0.68 W. Graph (d) shows the peak power of the output pulse as a function of the peak power of the input microwave pulse for three V_a values.

The data in Figure 5.17 show three important results. (1) The application of a positive voltage to the Pt element leads to an enhancement in the amplitude of the output signal and an increase in the power level to which the output pulse saturates. This indicates that when $V_a > 0$, the ISS effect leads to the amplification of the spin wave and plays a role of negative damping. (2) In contrast, the application of a negative voltage to the Pt element leads to a decrease in both the amplitude and saturation power level of the output pulse. This indicates that when $V_a < 0$, the ISS effect results in an attenuation

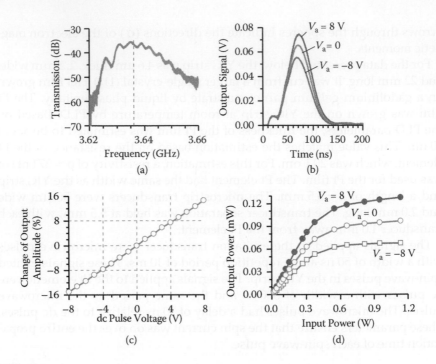

FIGURE 5.17

Control of spin waves through interfacial spin scattering. (a) Transmission profile for the trans-ducer-YIG-transducer structure. (b) Time-domain output signals for different dc pulse voltages (V_a) applied to the Pt layer. (c) Relative change in output signal amplitude as a function of V_a. (d) Output power versus input power responses for three V_a values, as indicated. (Reprinted with permission from Z. Wang, et al., *Phys. Rev. Lett.* 107, 146602 (2011). Copyright 2011 by the American Physical Society.)

in the spin-wave amplitude and plays a role of additional damping. (3) Over the highest available V_a range, the output signal amplitude versus V_a response shows almost perfect linear behavior. In addition, the data in (b) also indicates that a short pulse travels slightly faster than a tall pulse. This can be explained by the non-linearity-associated dependence of the spin-wave group velocity (v_g) on amplitude.[55]

These results can be interpreted as follows. When spin-polarized electrons scatter off the YIG surface, they transfer a certain net angular momentum to the surface spins in the YIG film. This momentum transfer is realized through the *s*-*d* exchange interactions at the Pt/YIG interface. Here, *s* refers to spin-polarized conduction electrons in the Pt layer, while *d* refers to localized electrons on the YIG surface. A theoretical model of this interfacial process has been suggested recently.[25]

The interfacial momentum transfer results in a net torque on the surface magnetic moments in the YIG film[43,44]:

$$\tau = C \frac{\gamma J_s}{M_s T} \hat{\mathbf{m}} \times \hat{\sigma} \times \hat{\mathbf{m}} \tag{5.17}$$

where $\hat{\mathbf{m}}$ is a unit vector along the magnetic moment direction of the YIG surface spins, γ is the absolute value of the gyromagnetic ratio, M_s is the saturation magnetization of the YIG film, T is the effective thickness of the YIG surface layer where the spins are involved in the momentum transfer, and C is a phenomenological coefficient that describes the properties of the Pt/YIG interface, such as the spin mixing conductance. J_s is the density of the spin current in the Pt layer, which can be written as

$$J_s = \theta_{SH} J_c = \theta_{SH} \frac{|V_a|}{\rho L} \tag{5.18}$$

where θ_{SH} and J_c are the spin Hall angle of and the charge current in the Pt element, respectively. The torque τ is exerted on the surface magnetic moments in the YIG film, but its effect is extended to other moments across the YIG thickness via dipolar and exchange interactions.

One can write the moment vector $\hat{\mathbf{m}}$ as $\mathbf{m}_0 + \mathbf{m}(t)$, where $\mathbf{m}_0$ is along the precession axis and is typically considered static, and $\mathbf{m}(t)$ is perpendicular to the precession axis and is dynamical. The magnitude of $\mathbf{m}(t)$ defines the spin-wave amplitude. If one takes the small-signal approximation, namely, $|\mathbf{m}(t)| \ll |\mathbf{m}_0|$, one can rewrite Equation (5.17) as

$$\tau = \mathrm{sgn}(-\mathbf{m}_0 \cdot \hat{\sigma}) C \frac{\gamma J_s}{M_s T} \mathbf{m}(t) \tag{5.19}$$

With this equation, one can qualitatively understand the results shown in Figure 5.17, as discussed below.

First, when $V_a > 0$, the vector $\hat{\sigma}$ is along the $-y$ direction, as shown in Figure 5.16, and the torque τ is parallel to the moment $\mathbf{m}(t)$ and tends to open up the precession cone. As a result, the spin-wave pulse is amplified and its saturation power level is raised up. Second, when $V_a < 0$, $\hat{\sigma}$ is along the $+y$ direction and τ is antiparallel to $\mathbf{m}(t)$. The net effect is that the spin-wave pulse is attenuated and saturates to a lower power level. Third, Equations (5.18) and (5.19) also indicate that $|\tau|$ increases with $|V_a|$. This explains the behavior shown in Figure 5.17(c).

The above interpretation was supported by measurements performed with the direction of $\hat{\sigma}$ fixed and the direction of $\mathbf{m}_0$ varied. The representative data are given in Figure 5.18. The figure shows $(V - V_0)/V_0$ as a function of V_a. The open circles show the same data as in Figure 5.17(c), which were measured for $H = 683$ Oe. The solid circles show the data for $H = -683$ Oe. This negative field supports the propagation of spin waves with a wave vector $\mathbf{k}$

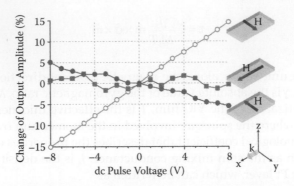

FIGURE 5.18

Relative change in output pulse amplitude as a function of dc pulse voltage applied to the Pt film for three different field orientations, as indicated. (Reprinted with permission from Z. Wang, et al., *Phys. Rev. Lett.* 107, 146602 (2011). Copyright 2011 by the American Physical Society.)

near the bottom surface of the YIG film. The squares show the data for a field of 1,074 Oe in the $+x$ direction. This field orientation supports the propagation of backward volume spin waves. For the volume wave measurements, the frequency of the input microwave pulse was 5.036 GHz. All other parameters were the same as cited above.

Three results are evident in Figure 5.18. (1) A reversal of the field orientation from the $+y$ direction to the $-y$ direction leads to opposite behavior, and this is true for both $V_a > 0$ and $V_a < 0$. This observation agrees with the expectation of Equation (5.19), namely, that a flip of $\mathbf{m}_0$ results in a switching between the $\boldsymbol{\tau}||\mathbf{m}$ and $\boldsymbol{\tau}||(-\mathbf{m})$ configurations and a corresponding switching between the positive damping and negative damping roles of $\boldsymbol{\tau}$. (2) For a field applied along the length of the YIG strip, the effects of V_a on V are insignificant. This agrees with the prediction of Equation (5.17), namely, that the net effect of $\boldsymbol{\tau}$ is zero over each precession period when $\mathbf{m}_0$ and $\hat{\boldsymbol{\sigma}}$ are normal to each other. (3) The ISS effects on the spin waves with larger amplitude near the bottom surface are weaker than those on the waves with larger amplitude near the top surface. This may indicate that the efficiency of transfer of the ISS effects on the surface spins to other spins decreases with film thickness.

Turn now to the evaluation of the ISS-produced changes in the decay rate η and damping constant α of the spin waves. One can use the following equation to model the propagation of a spin-wave pulse $u(x, t)$ along an YIG strip:

$$i\frac{\partial u}{\partial t} = \left[\omega_0 + v_g\left(-i\frac{\partial}{\partial x} - k_0\right)\right]u - i\eta u + i\Delta\eta u \tag{5.20}$$

where ω_0 and k_0 are the spin-wave carrier frequency and wave number, respectively. $\Delta\eta$ describes the ISS-produced change in η. It is clear from the above discussion that the sign of $\Delta\eta$ depends on $\text{sgn}(\mathbf{m}_0 \cdot \hat{\boldsymbol{\sigma}})$, and the magnitude of $\Delta\eta$ increases with $|\tau|$ (and thereby with $|V_a|$). To perform the quantitative calculation of $\Delta\eta$, one needs to know the spin mixing conductance at the YIG/Pt interface and consider the specific spin-wave configuration. For a qualitative discussion below, one introduces a phenomenological expression:

$$\Delta\eta = \text{sgn}(\mathbf{m}_0 \cdot \hat{\sigma})\lambda \frac{\theta_{SH}|V_a|}{\rho L} \qquad (5.21)$$

where the coefficient λ describes the efficiency of the change of η due to the ISS process. One can use Equation (5.20) to determine the spatial variation of the peak amplitude of the pulse $u(x, t)$. Assuming A_0 and A are the peak amplitudes of $u(x, t)$ at $x = L$ for $V_a = 0$ and $V_a \neq 0$, respectively, one then obtains

$$\frac{A - A_0}{A_0} = e^{\Delta\eta L/v_g} - 1 \qquad (5.22)$$

This ratio equals $(V - V_0)/V_0$. Thus, one can use the data in Figure 5.18 and Equation (5.22) to estimate $\Delta\eta$.

Figure 5.19 gives the ISS-produced changes in η and α for the experimental configurations, which are the same as for the data shown in Figures 5.17(b) and (c). The left and right axes show $\Delta\eta$ and $\Delta\alpha$, respectively, as a function of J_s. The estimation of $\Delta\alpha$ was based on the assumption that η was close to the decay rate of a uniform mode. The estimation used $2\Delta\eta = \Delta\alpha\gamma(2H + 4\pi M_s)$,[56] with $\gamma = 2.8\,\text{MHz/Oe}$ and $4\pi M_s = 1,750\,\text{G}$. The J_s values were obtained with Equation (5.18) and $\theta_{SH} = 0.076$. The data in Figure 5.19 clearly show that one can control, either enhance or mitigate, the decay of a spin wave through the ISS process.

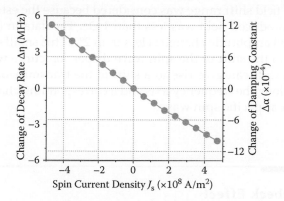

FIGURE 5.19
Change in spin-wave decay rate and damping constant as a function of spin current density in the Pt layer. (Reprinted with permission from Z. Wang, et al., *Phys. Rev. Lett.* 107, 146602 (2011). Copyright 2011 by the American Physical Society.)

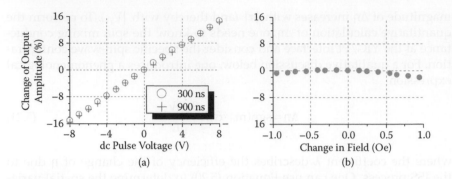

FIGURE 5.20
(a) Relative change in output pulse amplitude as a function of dc pulse voltage for two dc pulse durations. (b) Relative change in output pulse amplitude as a function of a shift in magnetic field. (Reprinted with permission from Z. Wang, et al., *Phys. Rev. Lett.* 107, 146602 (2011). Copyright 2011 by the American Physical Society.)

Fitting the $\Delta\eta$ values with Equation (5.21) yields $\lambda = 1.03 \times 10^{-2}$ m$^2 \cdot$ Hz/A for the $H||(+y)$ case and $\lambda = 0.33 \times 10^{-2}$ m$^2 \cdot$ Hz/A for the $H||(-y)$ case.

In addition to the spin current, the electric current in the Pt layer also produces ohmic heating and Oersted fields, but both have negligible influences on spin waves for the configurations in the present work. Figure 5.20 shows representative data. Graph (a) gives $(V - V_0)/V_0$ as a function of V_a. The circles give the same data as in Figure 5.17(c), for which the dc pulse width was 300 ns. The crosses give the data measured with a dc pulse width of 900 ns. One sees an almost perfect agreement between the two sets of data. This indicates that the heating effect is very weak. Graph (b) gives the relative change of the output amplitude as a function of a small shift in H. The data were measured for $V_a = 0$. Other parameters are the same as cited above. A ±1 Oe field shift range was considered because the estimated value of the current-produced field was less than 1 Oe. The data in (b) show that the field-induced amplitude change is less than 2%. This small change probably results from the shift of the spin-wave dispersion curve with the field. The dispersion shift can give rise to a shift of the transmission profile (see Figure 5.17(a)) along the frequency axis and a corresponding change in transmission loss for a specific spin wave.

5.5 Spin Seebeck Effect

5.5.1 Introduction

The spin Seebeck effect (SEE) was first observed in 2008 and refers to the generation of a spin voltage caused by a temperature gradient in magnetic materials, which allows the thermally induced injection of spin currents from the magnetic material into an attached nonmagnetic metal over several millimeters. An understanding of the spin Seebeck effect has developed from both theoretical and experimental aspects. It has been shown that the magnon and phonon degrees of freedom play crucial roles in the spin Seebeck effect. In this section, we present a brief review of the series of observations of the spin Seebeck effect in various magnetic materials, including magnetic metallic materials, magnetic insulators, and magnetic semiconductors, and also review the theoretical basis for understanding the spin Seebeck effect. Later, we will provide details on tuning the magnetization relaxation rate through the spin Seebeck effect.

The conventional Seebeck effect refers to the generation of electron flow induced by a thermal gradient across the conducting materials. This effect has been widely used in the field of thermal sensors. The spin Seebeck effect is the magnetic term of the conventional Seebeck effect, which is used to describe the generation of a spin voltage caused by a temperature gradient across magnetic materials, not necessarily conducting materials. The spin voltage difference across the magnetic materials can generate a spin current, which could be a flow of spin-polarized electrons or propagation of magnetization dynamic motions, named spin current. As shown in Figure 5.21, the conventional Seebeck effect produces electron current, and produces spin current. As indicated in Figure 5.21(b), the electrons carrying different spins will move in opposite directions when subjected to a temperature gradient, and overall a net spin current propagation along one direction is observed.

The spin Seebeck effect was first reported by K. Uchida in 2008.[19] The SSE was observed in a 20-nm-thick, soft ferromagnetic $Ni_{81}Fe_{19}$ film. In a metallic magnet, the spin-up and spin-down conduction electrons have different scattering rates and densities, and so the Seebeck coefficients (S) are different. Therefore, when a metallic magnet is subjected to a temperature gradient, it should generate different driving powers of electrons in different spin channels along the temperature gradient direction. This driving power of electrons generates differing amounts of flow in the two spin channels, that is, a spin current.

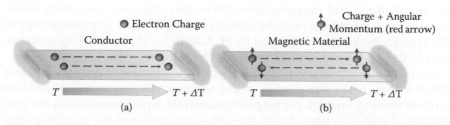

FIGURE 5.21
Graph (a) shows the conventional Seebeck effect. Graph (b) shows the spin Seebeck effect.

Regarding the ways to detect the spin current generated by SSE, one can utilize the inverse spin Hall effect (ISHE). ISHE refers to a phenomenon that a spin current can excite an electric current in a direction transverse to itself, which is the reverse of the Hall effect. It should also be noted that strong spin Hall and inverse spin Hall effects usually appear in materials with strong spin-orbit interactions, such as noble metals. Thin Pt layers are usually employed to probe spin current through the ISHE. As shown in Figure 5.6, the two NM layers on top of the magnetic material are the probes for spin current detection. As the spin current becomes established in the magnetic material along the temperature gradient direction, potential differences will be established in the NM layers. The amplitude of the induced potential is small, usually around the micron volt range. Usually a nanovoltmeter can be used to measure the potential difference. The difference $T_m(x) - T_p(x)$ gives rise to a spin current in the NM film element placed on the FM film strip. Such spin current can be expressed in the form of Equation (5.5). This spin current then produces an electric voltage across the NM film via the ISHE. One can expect a change in the sign of the electric voltage when the NM element is moved along the FM film strip from one end to another.

K. Uchida first reported the observation of the spin Seebeck effect in a metallic magnetic thin-film $Ni_{81}Fe_{19}$ in 2008. Then he reported the observation of SSE in a magnetic insulator $LaY_2Fe_5O_{12}$.[2] According to his report, the observation of the SSE in a magnetic insulator is realized by the ISHE in Pt films. It is similar to that described in his early work on $Ni_{81}Fe_{19}$ film, but the metallic magnetic metal layer is replaced by the garnet-type ferrimagnetic insulator $LaY_2Fe_5O_{12}$. This work demonstrates that conducting materials are not necessary for generation of spin current through the spin Seebeck effect.

In 2010, C. M. Jaworski reported the observation of the spin Seebeck effect in a ferromagnetic semiconductor, GaMnAs, which allows flexible design of the magnetization directions, a larger spin polarization, and measurements across the magnetic phase transition.[20] In his report, the SEE is observed even in the absence of longitudinal charge transport. The spatial distribution of spin currents is maintained across electrical breaks, indicating that the spin Seebeck effect has its local nature.

Later experimental works have repeated the above observations in different magnetic materials, including various magnetic metallic films, magnetic insulated films, and magnetic semiconductors. Theories have been developed to understand the observations, including a linear response theory of the spin Seebeck effect developed by H. Adachi[57] and phonon drag contribution to the spin Seebeck effect developed by J. Xiao.[21]

According to the linear response theory, the essence of the spin Seebeck effect is that the localized spins in the ferromagnet are excited by the heat current flowing through the magnetic materials, which then causes the generation of spin injections due to the imbalance between the pumping component and the backflow component spin current; a net spin current will be produced by the imbalance. It is important to know that the heat current that excites the localized spins actually has two contributions: one is the magnon heat current, and the other is the phonon heat current. The spin Seebeck effect also includes two processes: the first is called the magnon-driven spin Seebeck effect, in which the localized spins are thermally excited by the magnon heat current. For the second process, which is called the phonon drag spin Seebeck effect, the localized spins are excited by the phonon heat current. Phonon drag is a well-established idea in thermoelectricity. The concept of phonon drag proposes the idea that the nonequilibrium phonons driven by a temperature gradient can produce thermopower, which then drags electrons and induces the electron motions. According to the phonon drag theory, the role of nonequilibrium phonons is the key in the spin Seebeck effect. And the spin Seebeck signal due to phonon drag is proportional to the phonon lifetime.

The spin Seebeck effect has been demonstrated to be an important approach to inject spin current into nonmagnetic layers through a thermal gradient. Generation of spin current always associates with spin transfer torque between the magnetic materials and nonmagnetic materials, which will enable modification of the magnetization dynamics. Manipulating the magnetization relaxation rate with a thermal gradient is one of the most interesting functionalities of the spin Seebeck effect. The following section will introduce the experimental observation and underlying physics of tuning the ferromagnetic relaxation in magnetic thin films with the spin Seebeck effect.

In recent years, plenty of work has been done on tuning the magnetization relaxation rate utilizing spin-transfer torque. Basically, three approaches have been demonstrated that can control ferromagnetic relaxation in magnetic thin films.[42,43] The first approach makes use of the flow of spin-polarized electrons through the films.[58] The second takes advantage of the injection of spin-polarized electrons into the films. The third uses the scattering of spin-polarized electrons off the film surfaces.[59] Although these approaches differ in how they use spin-polarized electrons, they all rely on angular momentum transfers between the spin-polarized conduction electrons and the spins in the films to realize relaxation control. Equation (5.23) is the modified torque equation with the spin-transfer torque (STT) term appended. The STT term has a format similar to that of the damping term, which could perform as an additional damping or pumping force to drive the magnetization precession. The sign is determined by the cross product of the electron polarization and magnetization:

$$\frac{d\hat{m}}{dt} = -\gamma\hat{m}\times H + \alpha\hat{m}\times\frac{d\hat{m}}{dt} + C\hat{m}\times(\hat{\sigma}\times\hat{m}) \tag{5.23}$$

Figure 5.22 shows the schematic drawing that presents the functionality of the STT. The SST term works as a pumping force when the electron polarization is antiparallel with the magnetization. It works as a damping force when the electron polarization is parallel with the magnetization.

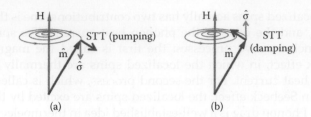

FIGURE 5.22
Schematic diagram of STT function. Graphs (a) and (b) show the situations for two difference STT functions.

5.5.2 Tuning of Magnetization Relaxation Rate through SSE

In 2012, L. Lu et al. demonstrated the tuning of the magnetization relaxation rate through the spin Seebeck effect,[60] which provides a new approach for relaxation control, specifically the control of spin-wave resonance linewidth in magnetic thin films through thermally induced interfacial spin transfers. In their experiments, a trilayered structure element that consists of a micron-thick YIG film grown on a sub-millimeter-thick GGG substrate and capped with a nanometer-thick Pt layer is used. The YIG film is ferromagnetic material, while the GGG substrate and the Pt layer are both paramagnetic materials.

A temperature gradient is established across the thickness of the GGG/YIG/Pt element by using two Peltier devices. The well-established temperature gradient produces, through the spin Seebeck effect, a spin current that flows from the YIG/Pt interface into the Pt layer.[20,61,62] The net effect of the spin current is an angular momentum transfer between the spins in the YIG film and the conduction electrons in the Pt layer, as shown in Figure 5.23. This momentum transfer results in a torque on the magnetic moments in the YIG film. The torque can either speed up or slow down the ferromagnetic relaxation in the YIG film, depending on the direction of the temperature gradient with respect to the trilayered structure. In their work, the control of the relaxation is presented in terms of changes in the linewidths of lateral spin-wave resonance modes in the YIG film.

Two points should be emphasized. (1) There exists a substantial difference between the approach demonstrated by L. Lu et al. and those demonstrated previously. Previous approaches all relied on external systems to supply spin-polarized electrons. In contrast, this approach has no need for an external supply of spin-polarized electrons, but requires the application of a temperature gradient. (2) The new approach is simple yet very efficient. As they reported, an easily accessible temperature gradient can produce a change in damping that is larger than the intrinsic damping in YIG materials.

Figure 5.24 shows the experimental configuration in their work. The core component is a GGG/YIG/Pt rectangular element. A temperature gradient

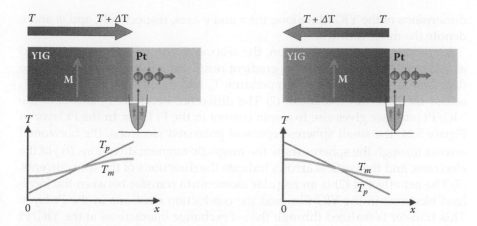

FIGURE 5.23
See color insert. (a) The phonon temperature is higher than the magnon temperature at the interface, which enables the spin injection with up-polarization from the magnetic layer to the NM layer. This process causes a pumping effect on magnetization precession in YIG. (b) The phonon temperature is lower than the magnon temperature at the interface, which enables the spin injection with down-polarization from the magnetic layer to the NM layer. This process causes a pumping effect on the magnetization precession in YIG.

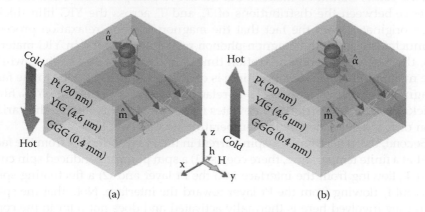

FIGURE 5.24
See color insert. Experimental configuration. Graphs (a) and (b) show the situations for two difference temperature gradients. (Reprinted with permission from L. Lu, et al., *Phys. Rev. Lett.* 108, 257202 (2012). Copyright 2012 by the American Physical Society.)

is applied across the thickness of the GGG/YIG/Pt element by placing it against two Peltier devices, which are not shown in Figure 5.24. An external field H is applied in the $+y$ direction to magnetize the YIG film. A microwave field h is applied along the x axis to excite spin waves in the YIG film. Due to the confinement of YIG lateral dimensions, the spin waves are standing modes with wave numbers $k = \sqrt{(m\pi/a)^2 + (n\pi/b)^2}$, where a and b are the dimensions of the YIG film along the x and y axes, respectively, and m and n denote the mode indexes.

According to their explanation, the relaxation control can be interpreted as follows. (1) The temperature gradient results in a difference between the distributions of the magnon temperature T_m and the phonon temperature T_p across the YIG film thickness. (2) The difference between T_m and T_p at the YIG/Pt interface gives rise to a spin current in the Pt layer. In the Pt layer in Figure 5.24, the small spheres represent polarized electrons, the horizontal arrows through the spheres show the magnetic moment directions ($\hat{\sigma}$) of the electrons, and the vertical arrows indicate the directions of the spin currents. (3) The net effect of (2) is an angular momentum transfer between the localized electrons in the YIG film and the conduction electrons in the Pt layer. This transfer is realized through the s-d exchange interactions at the YIG/Pt interface.[25] (4) The angular momentum transfer then produces a torque on the magnetic moments at the YIG surface and thereby affects the relaxation of those moments. (5) The effect of this spin-transfer torque (STT) is extended to other moments across the YIG film thickness via dipolar and exchange interactions. In the YIG layer in Figure 5.24, the longer arrows indicate magnetic moment directions ($\hat{m}$) and the shorter arrows indicate STT directions ($\hat{\tau}$).

Two points should be made about the above interpretation. First, the difference between the distributions of T_m and T_p across the YIG film thickness originates from the fact that the magnon-magnon relaxation process is much faster than the magnon-phonon relaxation process. In YIG materials, the magnon-magnon relaxation time is in the 10^{-9} to 10^{-7} s range, while the magnon-phonon relaxation time is on the order of 10^{-6} s. Due to the fast magnon-magnon relaxation, T_m is relatively constant across the YIG film thickness, and its distribution deviates from that of T_p.[61] Note that the variation of T_p is determined by the temperature gradient applied.

Second, the generation of a spin current in the Pt layer derives from the fact that at a finite temperature, there coexist (1) a spin pumping-induced spin current $\mathbf{I}_{sp}$ flowing from the interface into the Pt layer and (2) a fluctuating spin current $\mathbf{I}_{fl}$ flowing from the Pt layer toward the interface. Note that the spin pumping involved here is thermally activated and does not refer to the conventional spin pumping that is due to the application of external microwaves.[12] The magnitude of $\mathbf{I}_{sp}$ depends on T_m in the YIG layer near the interface, while that of $\mathbf{I}_{fl}$ depends on T_p in the Pt layer. Note that T_p in the Pt layer is essentially the same as T_p in the YIG layer near the interface because the Pt layer is very thin in comparison with the YIG film. The net spin current in the Pt layer is

$$\langle \mathbf{I} \rangle_y = \langle \mathbf{I}_{sp} \rangle_y + \langle \mathbf{I}_{fl} \rangle_y = \frac{2|\gamma|\hbar g_r k_B}{4\pi M_s V}(T_m - T_p) \qquad (5.24)$$

where $|\gamma|$ is the gyromagnetic ratio, g_r is the real part of the spin mixing conductance at the YIG/Pt interface, $4\pi M_s$ is the YIG saturation induction, and

V is the YIG volume in which the spins are involved in the interfacial spin transfer. One can see from Equation (5.24) that a difference between T_m and T_p results in a nonzero spin current in the Pt layer.

For the configuration in Figure 5.24(a), the temperature gradient results in $T_m > T_p$ at the YIG/Pt interface and $I_y > 0$ in the Pt layer. This configuration is similar to the conventional spin pumping effect for which the spin current consists of polarized electrons with magnetic moments antiparallel to the precession axis of the YIG magnetic moments, namely, $\hat{\sigma} // \langle -\hat{m} \rangle$. The net effect of this current is a torque on the YIG moments that enhances the relaxation. When the temperature gradient is reversed, as shown in Figure 5.24(b), one can expect $T_m < T_p$ at the interface and a spin current with $I_y < 0$ and $\hat{\sigma} // \langle \hat{m} \rangle$ in the Pt layer. In this case, the torque counters the relaxation and thereby plays a role of negative damping. It is important to emphasize that the asymmetry of the GGG/YIG/Pt structure is critical for the realization of the relaxation control. For a symmetric structure such as Pt/YIG/Pt, one might have opposite effects at the two YIG/Pt interfaces and an overall change of zero in the relaxation rate.

For the data presented below, the sample was prepared with three steps: (1) the growth of a 4.6-μm-thick YIG film on a 0.4-mm-thick GGG substrate by liquid phase epitaxy, (2) the growth of a 20-nm-thick Pt layer on the YIG film by pulsed laser deposition, and (3) the cutting of the GGG/YIG/Pt structure into a rectangular element with $a = 2.0$ mm and $b = 2.2$ mm. The temperatures of the two surfaces of the element were monitored by two thermocouples. In the discussions below, the temperatures at the top (Pt) and bottom (GGG) surfaces are referred to as T_1 and T_2, respectively. The spin-wave resonance measurements were carried out at 15 GHz in a shorted waveguide. Measurements were also carried out on a control sample GGG(0.4 mm)/YIG(4.6 μm)/Cu(20 nm) that had $a = 3.0$ mm and $b = 2.2$ mm. Two notes should be made about the data presented below. (1) All the linewidth data are peak-to-peak linewidths. (2) For the data points with error bars, the points show averaged values over six measurements and the bars give the corresponding standard deviations.

Figure 5.25 shows representative results for the GGG/YIG/Pt sample. Graph (a) shows a spin-wave resonance profile measured at $T_1 = T_2 = 30°C$. The integers show mode indexes (m, n) obtained on the basis of the spin-wave theory and the resonance fields. Graphs (b) and (c) show the linewidth and resonance field of mode (1, 1), respectively, as a function of $T_1 - T_2$. They were obtained with T_1 fixed at 30°C and T_2 varied from 10 to 50°C.

The data in Figure 5.25 show three results. (1) There exist a number of spin-wave resonance modes in the YIG film. In (a), the modes to the left and right of mode (1, 1) are usually classified as surface and backward volume modes, respectively.[10,63] (2) The linewidth of mode (1, 1) changes significantly with the temperature gradient. When the top surface is hot, the linewidth decreases with an increase in $|T_1 - T_2|$; when the top surface is cold, the linewidth increases with $|T_1 - T_2|$. These responses agree with the expectations. (3) The overall

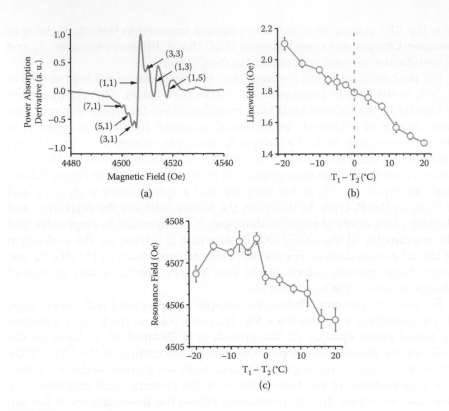

FIGURE 5.25

(a) Spin-wave resonance profile of a GGG/YIG/Pt element measured at $T_1 = T_2 = 30°C$. (b) Linewidth and (c) resonance field of mode (1, 1) as a function of $T_1 - T_2$. T_1 and T_2 denote the temperatures of the top (Pt) and bottom (GGG) surfaces, respectively. In (a), the integers indicate the indexes of resonance modes. The data in (b) and (c) were obtained with T_1 fixed at 30°C and T_2 varied from 10 to 50°C. (Reprinted with permission from L. Lu, et al., *Phys. Rev. Lett.* 108, 257202 (2012). Copyright 2012 by the American Physical Society.)

changes in linewidth and resonance field are ±17.8 and ±0.02%, respectively. These values indicate that the heating-associated resonance shift is insignificant and the observed change in linewidth is not due to a usual heating effect. As part of the sanity check, the same test was done on another GGG/YIG/Pt sample. The same effect was also observed in another GGG/YIG/Pt sample. The YIG was prepared in the same way, which also has a thickness of 4.6 μm and 0.4-mm-thick GGG as substrate. A 10-nm-thick Pt layer was deposited on the YIG film by pulsed laser deposition. The GGG/YIG/Pt element has a rectangular shape and a dimension of 2.0 × 1.8 mm. The sample was loaded into the same thermal device for the generation of a stable thermal gradient across the thickness of the whole structure. For this case, the spin-wave resonance measurements were carried out at 13 GHz in a shorted waveguide. This measurement is taken as a reproducibility check on the new phenomenon. Figure 5.26 shows the results for the second GGG/YIG/Pt sample. Graph (a)

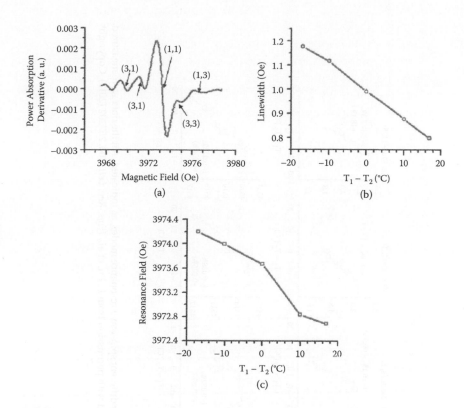

FIGURE 5.26
(a) Spin-wave resonance profile of the second GGG/YIG/Pt element measured at $T_1 = T_2 = 23.5°C$. (b) Linewidth and (c) resonance field of mode (1, 1) as a function of $T_1 - T_2$. T_1 and T_2 denote the temperatures of the top (Pt) and bottom (GGG) surfaces, respectively. In (a), the integers indicate the indexes of resonance modes. The data in (b) and (c) were obtained with T_1 fixed at 23.5°C and T_2 varied from 5.5 to 41.5°C.

shows a spin-wave resonance profile measured at $T_1 = T_2 = 23.5°C$. The integers show mode indexes (m, n) obtained on the basis of the spin-wave theory and the resonance fields. Graphs (b) and (c) show the linewidth and resonance field of mode (1, 1), respectively, as a function of $T_1 - T_2$. They were obtained with T_1 fixed at 23.5°C and T_2 varied from 5.5 to 41.5°C.

Measurements were carried out on a control sample GGG(0.4 mm)/YIG(4.6 µm)/Cu(20 nm) that had $a = 3.0$ mm and $b = 2.2$ mm. Two notes should be made about the data presented below. (1) All the linewidth data are peak-to-peak linewidths. (2) For the data points with error bars, the points show averaged values over six measurements and the bars give the corresponding standard deviations.

Figure 5.27 shows the data for other modes of the first GGG/YIG/Pt sample. In each panel, the top graph shows the linewidth versus $T_1 - T_2$ response, and the bottom graph shows the resonance field as a function of $T_1 - T_2$. The percentages in each graph give the range of the overall linewidth or field

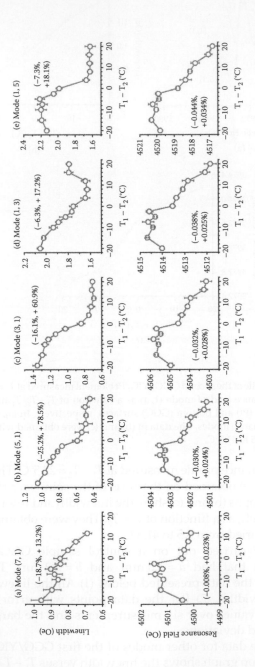

FIGURE 5.27

Linewidth (top) and resonance field (bottom) as a function of $T_1 - T_2$ for different spin-wave resonance modes, as indicated. All the data were measured with T_1 kept constant at 30°C and T_2 varied from 10 to 50°C. (Reprinted with permission from L. Lu, et al., *Phys. Rev. Lett.* 108, 257202 (2012). Copyright 2012 by the American Physical Society.)

change. Two important results are evident in Figure 5.27. First, except the top graph in (d), all the graphs show responses similar to those of mode (1, 1). Second, the surface modes show stronger STT effects than the backward volume modes. This agrees with the recent observation for spin pumping in YIG/Pt structures that surface modes showed significantly higher spin-pumping efficiencies than backward volume modes.[64] This behavior results from the difference in spin-wave amplitude distributions across the YIG film thickness for different modes.[63] Specifically, the amplitude for a surface spin wave is strong on one of the film surfaces and decays exponentially as one moves from the film surface into the film volume. In contrast, a backward volume mode has uniform amplitude across the film thickness. For the same microwave power applied, the surface modes have larger amplitudes near the YIG/Pt interface than the backward volume modes, and the net result is a larger spin transfer at the YIG/Pt interface and a larger change in linewidth. This is essentially the same as in conventional spin pumping, in which the magnitude of the spin current increases with the angle of the magnetization precession.

Figure 5.28 shows the data for other modes of the second GGG/YIG/Pt sample. Although less modes were observed for the second sample, similar

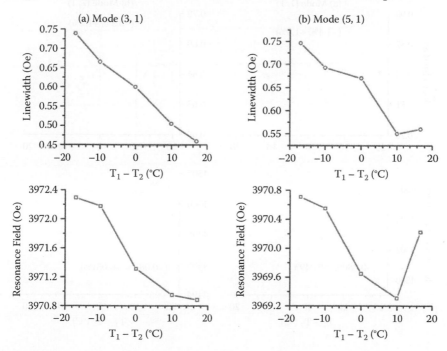

FIGURE 5.28
Linewidth (top) and resonance field (bottom) as a function of $T_1 - T_2$ for different spin-wave resonance modes, as indicated. All the data were measured with T_1 kept constant at 23.5°C and T_2 varied from 5.5 to 41.5°C.

effects were obtained. The data in Figures 5.27 and 5.28 together demonstrate the feasibility of resonance linewidth control through thermally induced spin transfers. For such control, both the temperature gradient and the Pt layer play crucial roles: the former gives rise to a spin current, and the latter acts as a sink to dissipate the spin current. Such roles were clearly demonstrated by additional measurements with different configurations and different samples.

Figure 5.29 shows data for the GGG/YIG/Cu sample. One can see that with a change in $T_1 - T_2$, the resonance field changes in the same manner as the GGG/YIG/Pt sample does, while the linewidth exhibits no notable changes. These responses result from the fact that the Cu layer cannot act as a spin sink because Cu has a long electron mean free path λ_f, a long spin-flip length λ_{sf}, and weak spin-orbit coupling. The spin diffusion length in Cu is $l_{sd} = \sqrt{(1/3)\lambda_f \lambda_{sf}} \approx 500$ nm.[65] This length is significantly larger than the thickness of the Cu layer in the GGG/YIG/Cu sample. In contrast, the spin diffusion length in Pt is only about 10 nm,[47] which is smaller than the thickness of the Pt layer in the GGG/YIG/Pt sample.

Figure 5.30 shows data for the first GGG/YIG/Pt sample measured with $T_1 = T_2$. The data show two results. (1) The resonance field increases

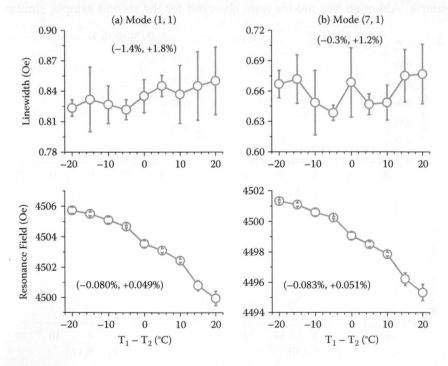

FIGURE 5.29
Linewidth (top) and resonance field (bottom) as a function of $T_1 - T_2$ for two different modes. The data were measured for a GGG/YIG/Cu sample with $T_1 = 30°C$ and $T_2 = 10 \sim 50°C$. (Reprinted with permission from L. Lu, et al., *Phys. Rev. Lett.* 108, 257202 (2012). Copyright 2012 by the American Physical Society.)

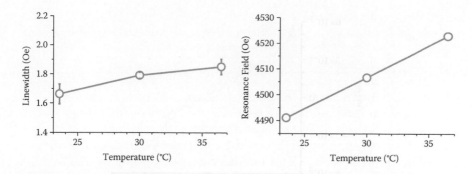

FIGURE 5.30
Linewidth and resonance field of mode (1, 1) as a function of the sample temperature for the
GGG/YIG/Pt sample. (Reprinted with permission from L. Lu, et al., *Phys. Rev. Lett.* 108, 257202
(2012). Copyright 2012 by the American Physical Society.)

significantly with the sample temperature, with an overall change of larger
than 0.6%. This response originates from the fact that $4\pi M_s$ decreases with
temperature. (2) Although the field changes are one order of magnitude larger
than the changes shown in Figures 5.27 and 5.28, the changes in linewidth
are smaller than those shown in Figures 5.27 and 5.28. Similar results were
obtained for the GGG/YIG/Cu sample. These results clearly demonstrate that
the linewidth changes in Figures 5.27 and 5.28 do not result from the heating
or cooling of the YIG film. Rather, they result from the temperature gradient.

Similar results were also obtained for the second GGG/YIG/Pt sample,
which is not shown.

One can use the measured linewidth changes to estimate the STT-produced
changes in damping, $\Delta\alpha_{STT}$, if one assumes that (1) the STT-produced damping is Gilbert-like and (2) the spin-wave linewidth is close to the linewidth of
a uniform mode. Figure 5.31 shows the $\Delta\alpha_{STT}$ values estimated with the data
in Figure 5.25(b). One can see that a temperature gradient of 20°C gives rise
to a change in damping of about 5.1×10^{-5}. The largest linewidth change was
observed for mode (3, 1). This change corresponds to a $\Delta\alpha_{STT}$ value of about
8.3×10^{-5}. These changes are substantial, as they are larger than the intrinsic
damping in YIG materials, which is $\alpha_0 = 3 \times 10^{-5}$. On the other hand, $\Delta\alpha_{STT}$ versus temperature difference of the second GGG/YIG/Pt sample is presented
in Figure 5.31. Similar to the data of the first sample, a temperature gradient
of 17°C gives rise to a change in damping of about 3.7×10^{-5}. Note that all the
measurements in this work were carried out with magnetic fields applied in
the film planes. Future work on the determination of $\Delta\alpha_{STT}$ through measurements with out-of-plane fields at different frequencies is of great interest.

Two points should be emphasized. (1) It is the overall temperature gradient across the GGG/YIG/Pt structure, rather than the gradient across the
YIG film, that is responsible for the demonstrated effects. This is because the
phonons have long propagation length and the magnons in the YIG film can
feel the temperature in the GGG substrate.[66] (2) Although the demonstrations

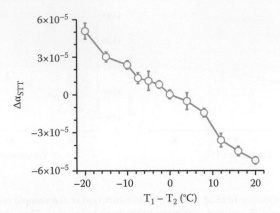

FIGURE 5.31
STT-produced change in damping constant α as a function of $T_1 - T_2$. The $\Delta\alpha_{STT}$ values were obtained from the data shown in Figure 5.25(b). (Reprinted with permission from L. Lu, et al., *Phys. Rev. Lett.* 108, 257202 (2012). Copyright 2012 by the American Physical Society.)

made use of YIG-based structures, one can expect similar effects in structures consisting of ferromagnetic metallic films. Also, it should be mentioned that the growth of a Pt layer on a YIG film does lead to a certain increase in the damping of the YIG film. Future work on the study of physical mechanisms for this increase and possible methods for avoiding it are of great interest.

Overall, Lu's work demonstrated the control of linewidths of spin-wave resonance modes in a GGG/YIG/Pt structure through the application of a temperature gradient across the structure thickness. Such control relies on the thermally induced spin angular momentum transfers across the YIG/Pt interface. The spin relaxation rate can be either enhanced or reduced by switching the thermal gradient direction, and the tuning rate is linearly dependent on the thermal gradient amplitude. The results not only demonstrate a new approach for relaxation control, but also suggest a new mechanism for spin torque oscillators, in which the spin torque results from thermally induced angular momentum transfer.

Plenty of work has been done in the field of the spin Seebeck effect. And controlling the magnetization relaxation rate is one of the promising applications. The whole field is growing very fast at present, it is still early, and there still exist many issues that remain unclear and need to be resolved.

5.6 Summary

This chapter provided a brief review of the rising field of spintronics on a magnetic insulator. After the introduction, a great amount of experimental results was presented along with detailed interpretations in the categories of spin pumping, magnetic proximity, spin Seebeck, and spin Hall effects. The discussions reached five important conclusions:

1. Spin pumping occurs at the YIG/NM interfaces. The efficiency of angular momentum transfer due to the spin pumping effect was studied in terms of spin mixing conductance. Thanks to the successful fabrication of high-quality YIG nanofilms by PLD, the spin pumping experiments give a spin mixing conductance of $\sim 10^{14}$ cm^{-2} for YIG/NM interfaces, which is about 10 to 15% of that for Fe/NM. Surface treatment by ion etching can significantly enhance the spin pumping effect at YIG/NM interfaces.

2. There exists a new damping that originates from the dynamic exchange coupling between the spins of ferromagnetic-ordered Pt atoms and the YIG spins. The FM ordering of Pt is caused by the magnetic proximity effect at the interface. This coupling results in the transfer of Pt damping to YIG, and thereby the enhanced damping. Fortunately, such new damping can be suppressed by simply inserting a Cu spacer.

3. Electric control of magnetization relaxation in YIG thin films via the ISS process was discovered. It was found that the ISS effect can play a positive or negative role in the relaxation, and one can control this role by simply changing the strength and direction of the dc current.

4. The amplification of the spin wave through the spin Hall effect is also presented. As the spin current scatters off the surface of the YIG film, it exerts a torque on the YIG surface spins. Due to the dipolar and exchange interactions, the effect of this torque is extended to other spins across the YIG thickness and thereby to spin-wave pulses that travel in the YIG film.

5. The tuning of the magnetization relaxation rate through the spin Seebeck effect has been demonstrated, which provides a new approach for relaxation control, specifically the control of spin-wave resonance linewidth in magnetic thin films through thermally induced interfacial spin transfers.

References

1. Y. Kajiwara, K. Harii, S. Takahashi, J. Ohe, K. Uchida, M. Mizuguchi, H. Umezawa, H. Kawai, K. Ando, K. Takanashi, S. Maekawa, and E. Saitoh, *Nature* 464, 262 (2010).
2. K. Uchida, J. Xiao, H. Adachi, J. Ohe, S. Takahashi, J. Ieda, T. Ota, Y. Kajiwara, H. Umezawa, H. Kawai, G. E. W. Bauer, S. Maekawa, and E. Saitoh, *Nat. Mater.* 9, 894 (2010).

3. K. Uchida, H. Adachi, T. Ota, H. Nakayama, S. Maekawa, and E. Saitoh, *Appl. Phys. Lett.* 97, 172505 (2010).
4. P. E. Wigen, R. D. McMichael, and C. Jayaprakash, *J. Magn. Magn. Mater.* 84, 237 (1990).
5. M. Wilson, *Physics Today*, May 2010, p. 13.
6. J. Sinova, *Nat. Mater.* 9, 880 (2010).
7. B. Lax and K. J. Button, *Microwave Ferrites and Ferrimagnetics* (McGraw-Hill, New York, 1962).
8. P. Kabos and V. S. Stalmachov, *Magnetostatic Waves and Their Applications* (Chapman & Hall, London, 1994).
9. J. D. Adam, L. E. Davis, G. F. Dionne, E. F. Schloemann, and S. N. Stitzer, *IEEE Trans. Microwave Theory Tech.* 50, 721 (2002).
10. D. Stancil and A. Prahakar, *Spin Waves—Theory and Applications* (Springer, New York, 2009).
11. Y. Tserkovnyak, A. Brataas, and G. E. W. Bauer, *Phys. Rev. B* 66, 224403 (2002).
12. B. Heinrich and J. A. C. Bland, *Ultrathin Magnetic Structures: Fundamentals of Nanomagnetism* (Springer, Berlin, 2005).
13. J. D. Jackson, *Classical Electrodynamics* (John Wiley & Sons, Danvers, MA, 1999).
14. S. Takahashi and S. Maekawa, *Sci. Technol. Adv. Mater.* 9, 014105 (2008).
15. M. I. Dyakonov and V. I. Perel, *JETP Lett.* 13, 467 (1971).
16. J. Hirsch, *Phys. Rev. Lett.* 83, 1834 (1999).
17. C. Day, *Phys. Today* 58(2), 17 (2005).
18. E. Saitoh, M. Ueda, and H. Miyajima, *Appl. Phys. Lett.* 88, 182509 (2006).
19. K. Uchida, S. Takahashi, K. Harii, J. Ieda, W. Koshibae, K. Ando, S. Maekawa, and E. Saitoh, *Nature* 455, 778 (2008).
20. C. M. Jaworski, J. Yang, S. Mack, D. D. Awschalom, R. C. Myers, and J. P. Heremans, *Phys. Rev. Lett.* 106, 186601 (2011).
21. J. Xiao, G. E. W. Bauer, K. Uchida, E. Saitoh, and S. Maekawa, *Phys. Rev. B* 81, 214418 (2010).
22. Y. Tserkovnyak, A. Brataas, and G. E. W. Bauer, *Phys. Rev. Lett.* 88, 117601 (2002).
23. B. Heinrich, Y. Tserkovnyak, G. Woltersdorf, A. Brataas, R. Urban, and G. E. W. Bauer, *Phys. Rev. Lett.* 90, 187601 (2003).
24. B. Kardasz, O. Mosendz, B. Heinrich, Z. Liu, and M. Freeman, *J. Appl. Phys.* 103, 07C509 (2008).
25. S. Takahashi, E. Saitoh, and S. Maekawa, *J. Phys. Conf. Ser.* 200, 062030 (2009).
26. Y. Sun, Y.-Y. Song, H. Chang, M. Kabatek, M. Jantz, W. Schneider, M. Wu, H. Schultheiss, and A. Hoffmann, *Appl. Phys. Lett.* 101, 152405 (2012).
27. B. Heinrich, C. Burrowes, E. Montoya, B. Kardasz, E. Girt, Y.-Y. Song, Y. Sun, and M. Wu, *Phys. Rev. Lett.* 107, 066604 (2011).
28. C. Burrowes, B. Heinrich, B. Kardasz, E. A. Montoya, E. Girt, Y. Sun, Y.-Y. Song, and M. Wu, *Appl. Phys. Lett.* 100, 092403 (2012).
29. R. E. Parra and R. Medina. *Phys. Rev. B* 22, 5460 (1980).
30. C. Liu and S. D. Bader, *Phys. Rev. B* 44, 2205 (1991).
31. K. J. Strandburg, D. W. Hall, C. Liu, and S. D. Bader, *Phys. Rev. B* 46, 10818 (1992).
32. P. Poulopoulos, F. Wilhelm, H. Wende, G. Ceballos, K. Baberschke, D. Benea, H. Ebert, M. Angelakeris, N. K. Flevaris, A. Rogalev, and N. B. Brookes, *J. Appl. Phys.* 89, 3874 (2001).
33. F. Wilhelm, P. Poulopoulos, H. Wende, A. Scherz, K. Baberschke, M. Angelakeris, N. K. Flevaris, and A. Rogalev, *Phys. Rev. Lett.* 87, 207202 (2001).

34. F. Wilhelm, M. Angelakeris, N. Jaouen, P. Poulopoulos, E. Th. Papaioannou, Ch. Mueller, P. Fumagalli, A. Rogalev, and N. K. Flevaris, *Phys. Rev. B* 69, 220404(R) (2004).
35. H. Wende, *Rep. Prog. Phys.* 67, 2105 (2004).
36. Y. M. Lu, Y. Choi, C. M. Ortega, X. M. Cheng, J. W. Cai, S. Y. Huang, L. Sun, and C. L. Chien, *Phys. Rev. Lett.* 110, 147207 (2013).
37. S. Visnovsky, M. Nyvlt, V. Prosser, R. Lopusnik, R. Urban, J. Ferre, G. Penissard, D. Renard, and R. Krishnan, *Phys. Rev. B* 52, 1090 (1995).
38. L. Uba, S. Uba, V. N. Antonov, A. N. Yaresko, T. Slezak, and J. Korecki, *Phys. Rev. B*, 62, 13731 (2000).
39. Y. Sun, H. Chang, M. Kabatek, Y.-Y. Song, Z. Wang, M. Jantz, W. Schneider, M. Wu, E. Montoya, B. Kardasz, B. Heinrich, S. G. E. teVelthuis, H. Schulthesis, and A. Hoffmann, *Phys. Rev. Lett.* 111, 106601, 2013.
40. M. Sparks, *Ferromagnetic-Relaxation Theory* (McGraw-Hill, New York, 1964).
41. G. D. Fuchs, J. C. Sankey, V. S. Pribiag, L. Qian, P. M. Braganca, A. G. F. Garcia, E. M. Ryan, Z. P. Li, O. Ozatay, D. C. Ralph, and R. A. Buhrman, *Appl. Phys. Lett.* 91, 062507 (2007).
42. D. C. Ralph and M. D. Stiles, *J. Magn. Magn. Mater.* 320, 1190 (2008).
43. K. Ando, S. Takahashi, K. Harii, K. Sasage, J. Ieda, S. Maekawa, and E. Saitoh, *Phys. Rev. Lett.* 101, 036601 (2008).
44. L. Liu, T. Moriyama, D. C. Ralph, and R. A. Buhrman, *Phys. Rev. Lett.* 106, 036601 (2011).
45. S. S. Kalarickal, P. Krivosik, J. Das, K. S. Kim, and C. E. Patton, *Phys. Rev. B* 77, 054428 (2008).
46. Y. Niimi, M. Morota, D. H. Wei, C. Deranlot, M. Basletic, A. Hamzic, A. Fert, and Y. Otani, *Phys. Rev. Lett.* 106, 126601 (2011).
47. L. Vila, T. Kimura, and Y. Otani, *Phys. Rev. Lett.* 99, 226604 (2007).
48. S. Bance, T. Schrefl, G. Hrkac, A. Goncharov, D. A. Allwood, and J. Dean, *J. Appl. Phys.* 103, 07E735 (2008).
49. T. Schneider, A. A. Serga, B. Leven, B. Hillebrands, R. L. Stamps, and M. P. Kostylev, *Appl. Phys. Lett.* 92, 022505 (2008).
50. A. Khitun, M. Bao, and K. L. Wang, *J. Phys. D* 43, 264005 (2010).
51. A. G. Gurevich and G. A. Melkov, *Magnetization Oscillations and Waves* (CRC Press, New York, 1996).
52. P. A. Kolodin, P. Kabos, C. E. Patton, B. A. Kalinikos, N. G. Kovshikov, and M. P. Kostylev, *Phys. Rev. Lett.* 80, 1976 (1998).
53. A.V. Bagada, G. A. Melkov, A. A. Serga, and A. N. Slavin, *Phys. Rev. Lett.* 79, 2137 (1997).
54. M. M. Scott, C. E. Patton, M. P. Kostylev, and B. A. Kalinikos, *J. Appl. Phys.* 95, 6294 (2004).
55. A. N. Slavin and C. E. Zaspel, *J. Appl. Phys.* 91, 8673(2002).
56. D. J. Craik, *Magnetic Oxide* (John Wiley, London, 1975).
57. H. Adachi, J.I. Ohe, S. Takahashi, and S. Maekawa, *Phys. Rev. B* 83, 094410 (2011).
58. Z. Wang, Y. Sun, M. Wu, V. Tiberkevich, and A, Slavin, *Phys. Rev. Lett.* 107, 146602 (2011).
59. Z. Wang, Y. Sun, Y. Song, M. Wu, H. Schultheiß, J. E. Pearson, and A. Hoffmann, *Appl. Phys. Lett.* 99, 162511 (2011).
60. L. Lu, Y. Sun, M. Jantz, and M. Wu, *Phys. Rev. Lett.* 108, 257202 (2012).

61. K. Uchida, T. Ota, K. Harii, S. Takahashi, S. Maekawa, Y. Fujikawa, and E. Saitoh, *Solid State Commun.* 150, 524 (2010).
62. E. Padrón-Hernández, A. Azevedo, and S. M. Rezende, *Phys. Rev. Lett.* 107, 197203 (2011).
63. R. W. Damon and J. R. Eshbach, *J. Phys. Chem. Solids* 19, 308 (1961).
64. C. Sandweg, Y. Kajiwara, K. Ando, E. Saitoh, and B. Hillebrands, *Appl. Phys. Lett.* 97, 252504 (2010).
65. T. Kimura, J. Hamrle, and Y. Otani, *Phys. Rev. B* 72, 014461 (2005).
66. K. Uchida, H. Adachi, T. An, T. Ota, M. Toda, B. Hillebrands, S. Maekawa, and E. Saitoh, *Nat. Mater.* 10, 737 (2011).

6

Electric Field-Induced Switching for Magnetic Memory Devices

Pedram Khalili and Kang L. Wang

Department of Electrical Engineering, University of California, Los Angeles, Los Angeles, California

CONTENTS

6.1 Introduction

The field of magnetism has witnessed a remarkable series of discoveries over the past few decades, and has given rise to new mechanisms to enable interactions between electrical and magnetic properties of materials and devices at the nanoscale. This has resulted in the emergence of various spintronic devices, where electrical currents and voltages can directly interact with the magnetization in nanostructures [1–10]. The discoveries of giant magnetoresistance (GMR) [8, 9] and tunneling magnetoresistance (TMR) [3–7] effects allowed for electrical reading of the relative magnetization orientations of different layers in spin valves and magnetic tunnel junctions (MTJs), respectively. This resulted in major advances in the field of magnetic sensors [1, 11–15], with applications in hard disk drives, magnetoresistive random access memory (MRAM), and biomedical sensing devices. New spintronic devices were further developed over the past decade, following the theoretical prediction and experimental realization of the spin-transfer torque (STT) effect [16–20], which allows for manipulation and switching of magnetic moments using spin-polarized electric currents. The STT effect offers many potential applications, most notably in random access memory (STT-MRAM) [2, 10, 19, 21–28] and microwave nano-oscillators [18, 20, 22, 29–37]. The appeal of STT for use in MRAM arises from the fact that unlike previous generations of MRAM, where bits are switched by magnetic fields induced by currents, in the case of STT-MRAM, writing is performed by passing currents directly through the MTJ memory bit. This results in advantages in terms of energy efficiency, density, and scalability over field-switched toggle MRAM. In addition, when combined with complementary metal-oxide silicon (CMOS) logic circuits, STT devices also offer the potential for the realization of nonvolatile and programmable logic circuits [2, 21, 38–40]. These circuits take advantage of the fact that standby power can be eliminated and intermediate computation steps are stored in a nonvolatile manner using the integrated nonvolatile memory, hence allowing for instant on/off capability. However, as will be outlined below, these logic solutions suffer from limitations due to the dynamic (i.e., switching) energy dissipation of STT devices.

Despite its significant potential as a nonvolatile memory candidate for replacing or complementing existing memory technologies such as dynamic and static random access memory (DRAM and SRAM), STT-MRAM still suffers from a shortcoming in terms of energy efficiency. This is a result of the need for driving relatively large electric currents through the device for switching, which, along with the nonzero voltage drop across the resistive MTJ bit, can lead to significant power dissipation. The lowest reported switching (i.e., writing) energy of STT-MRAM devices is currently on the order of ~100 fJ [23, 26, 27, 41–44], which is still more than two orders of magnitude larger than modern CMOS transistors, which consume <1 fJ per switch at scaled technology nodes. While this may not be a limitation for a number of memory applications where energy efficiency is not the main concern (e.g., where frequent switching operations are not needed), it can be an important issue in applications where memory is closely integrated with logic, hence necessitating frequent read and write operations. Examples of such applications include memories for systems on a chip [26] and hybrid

CMOS-MTJ nonvolatile logic circuits [38–40, 45, 46], which are significantly constrained by their energy consumption if STT-based devices are used for the memory [38]. As a result, and given that no other existing or emerging (nonmagnetic) nonvolatile memory technologies can reach the required energy efficiency for such applications either, there is currently great interest in alternative mechanisms for magnetic memory technologies that would enable a significant reduction in dynamic power dissipation.

A number of proposed solutions have been set forth to address this challenge. Some of them involve the use of new types of spin-transfer torques, which rely on the large spin-orbit interaction in nonmagnetic metals placed adjacent to a magnetic medium [47, 48]. As an example of this approach, we will briefly review devices that use the spin Hall effect (SHE) for switching in Section 6.3. More fundamentally, however, a dramatic reduction in energy dissipation in spintronic devices would involve a move from current-controlled to voltage-controlled mechanisms for magnetization switching. In this chapter we will review recent efforts to use voltages (i.e., electric fields) rather than currents to control magnetization. The use of electric fields allows for much lower power dissipation because in principle, no charge flow is required for such voltage-controlled devices to operate.

Several different approaches have been proposed for the control of magnetization using electric fields. These include using single-phase multiferroic materials [49, 50], multiferroic tunnel junctions [51], multiferroic heterostructures where magnetic and electrical properties are coupled through strain [52, 53], dilute magnetic semiconductors where carrier-mediated ferromagnetism is controlled by a gate voltage [54], and ultrathin metallic structures where ferromagnetic phase transition (i.e., Curie temperature) is controlled by an electric field [55, 56]. A particularly promising approach, which has been the subject of recent experimental and theoretical studies, relies on the electric field control of interfacial magnetic anisotropy [57–64]. The practical promise of this voltage-controlled magnetic anisotropy (VCMA) effect relies in part on the fact that it can be realized in materials very similar to those used in regular magnetic tunnel junctions, such as those used in STT devices. Hence, VCMA devices can benefit from the significant existing body of knowledge on materials and manufacturing of MRAM devices. While only a subset of the entire body of research on electric field control of magnetism, recent developments on VCMA devices point to the possibility of using these devices in practical MRAM chips for memory and logic applications, and hence VCMA devices will be the primary focus of this chapter.

The chapter is organized as follows. Section 6.2 presents a brief overview of the requirements for emerging nonvolatile memory technologies in order to be competitive compared to existing solutions, with a particular emphasis on embedded memories complementing CMOS logic circuits. This will be followed by an introduction to STT-MRAM devices in Section 6.3. While not the main focus of this chapter, the discussion of STT-MRAM will be used to illustrate some of the main trade-offs and performance metrics that are common to most MRAMs, before proceeding to the discussion of VCMA-based magnetoelectric RAM (MeRAM). Section 6.4 presents an introduction to the voltage-controlled magnetic anisotropy (VCMA) effect in layered magnetic material structures, which forms the basis for device applications discussed in later sections. Recent experimental results on interfacial perpendicular magnetic anisotropy (PMA) in magnetic devices, and in particular the modulation of PMA by electric fields, are discussed. Section 6.5 discusses microwave manifestations of VCMA, as observed using ferromagnetic resonance (FMR) measurements. Section 6.6 discusses applications of the VCMA effect in switching of magnetization in nanoscale devices, and its implications for magnetic memory devices. Several different recent experimental results are discussed and compared, including voltage-assisted thermally activated switching of magnetic memory bits, ultrafast precessional switching in both in-plane and perpendicularly magnetized devices, and hybrid devices such as spintronic charge trap memory cells. Section 6.7 presents a single-domain model of VCMA-induced switching in nanomagnets, which can be applied to analyze device-level measurement data and scaling trends for MeRAM. Section 6.8 focuses on circuit implementations of MeRAM based on VCMA-switched memory bits. In particular, there is a discussion of how the unique unipolar switching characteristics of VCMA devices can enable very high-density crossbar implementations, which significantly surpass the area efficiency of other MRAM solutions. This is followed by a summary and conclusions in Section 6.9.

6.2 Nonvolatile Spintronic Memory:
Requirements and Solutions

The continuous scaling of complementary metal-oxide silicon (CMOS) transistors over the past several decades has resulted in faster and more powerful microprocessors, and electronics with ever-increasing functionalities [65, 66]. While CMOS is currently the only solution for mainstream logic circuit implementations, it is quickly approaching its scaling limits due to increased power dissipation issues at scaled technology nodes. In particular, while scaling generally reduces the dynamic power dissipation of CMOS devices, the current scaling dilemma is a result of the increase in static (i.e., standby) leakage power, as well as from the increase in device density as the transistor size is scaled down. The integration of a fast, energy-efficient nonvolatile memory technology with CMOS logic can alleviate this problem. Hybrid nonvolatile logic circuits consisting of volatile CMOS and nonvolatile magnetic memory can thus allow for continued scaling with improved energy efficiency, by eliminating or substantially reducing the static power dissipation [21].

In current electronic systems, during computation information is often temporally stored in static random access memory (SRAM), which acts as cache memory placed close to the CMOS logic (often on the same chip), as well as in dynamic random access memory (DRAM) acting as the principal working memory (with higher density, and hence larger memory chip capacities than SRAM), and is then permanently stored in NAND Flash (e.g., in solid-state drive (SSD) storage), or hard disk drives (HDDs).

Table 6.1 compares the most important performance parameters for SRAM, DRAM, and NAND Flash (i.e., existing memory technologies) to STT-MRAM and MeRAM. Among the established technologies, SRAM is the fastest (operating in the GHz range), needs fairly little power (~100 fJ per switch), and has virtually unlimited endurance. However, SRAM is volatile, has a very low density (i.e., large cell size), and its standby power consumption becomes a problem, especially as the transistor dimensions are scaled down, leading to higher leakage currents. DRAM has a much higher density than SRAM, but is also volatile and needs a periodic refresh current, which results in significant power consumption overhead. NAND Flash has the highest density among the three and is nonvolatile. However, it is extremely slow and offers very limited endurance, restricting its application space to storage with infrequent access only. It also has the largest energy consumption per bit, and unlike the other technologies, large voltages are needed for its operation.

Spintronic devices (i.e., various types of MRAM) are strong candidates for nonvolatile memory due to the inherent hysteresis present in ferromagnetic materials, and the compatibility of their material stacks with standard CMOS processing [19, 25, 67]. Magnetoresistive RAM (MRAM) exhibits significant

TABLE 6.1

Comparison of Existing (Nonmagnetic) and Emerging Spintronic Memory Technologies, Highlighting STT-MRAM (Using Current-Induced Switching) and MeRAM (Using Electric Field-Controlled Switching)

Technology	SRAM	DRAM	NAND Flash	STT-MRAM	MeRAM
Energy/bit (fJ)	100	> 100	10^6	100	< 10
Write speed (ns)	1	20	10^6	1–10	1–10
Read speed (ns)	1	30	50	1	1
Cell size (F^2)	> 40	6–10	4	8–30	4–8
Endurance (cycles)	10^{16}	10^{16}	10^5	10^{16}	10^{16}
Nonvolatile	No	No	Yes	Yes	Yes
Standby power	Leakage	Refresh	None	None	None
Nonvolatile logic capability	No	No	No	Limited	Yes

Source: K. L. Wang, et al., *Journal of Physics D: Applied Physics*, 46, 074003, 2013.

advantages as a fast and high-endurance nonvolatile memory, which can be integrated with CMOS in a back-end-of-line (BEOL) process [23]. Readout of MRAM is accomplished via the tunneling magnetoresistance (TMR) [4] effect. For the writing of information, the first generations of MRAM utilized Oersted fields generated by running currents in adjacent conducting lines to switch the magnetization of memory cells. However, this strategy faces limitations in terms of energy efficiency, scalability, and density due to the large currents needed to accomplish the switching. The manipulation of magnetic moments by currents or electric voltages overcomes the shortcomings of magnetic field-switched MRAM.

In the next sections, we will first concentrate on the switching of MRAM bits using spin-polarized currents via the spin-transfer torque (STT) effect [16–18]. As seen in Table 6.1, the use of spin-polarized currents for STT-MRAM allows for switching energies and speed close to SRAM, while the density can be better than SRAM, making STT-MRAM a suitable candidate for a number of embedded applications. The focus of this chapter will be on mechanisms beyond STT, targeting ultra-low-power performance (switching energies of <10 fJ and below). As shown in Table 6.1, MeRAM (i.e., electric field-controlled MRAM) is the only memory technology that can potentially reach such low energies. In addition, as will be outlined in Section 6.8, the use of electric field-controlled devices also provides a distinct advantage in terms of cell size (i.e., memory density), bringing MeRAM close to NAND Flash densities. In terms of power efficiency, dynamic switching energies of STT-MRAM are still around two orders of magnitude higher than those of CMOS (~1 fJ per switch for the 32 nm node), limiting the possibility of integration of STT devices at the gate level with CMOS logic for nonvolatile logic applications [21], where frequent access of the memory device is needed. Hence, ultra-low-power voltage-controlled MeRAM can impart a significant overall energy advantage over STT-based circuits by offering the lowest dynamic energy dissipation of any nonvolatile memory, as well as the elimination of standby power.

For ultra-low-power spin-based memories beyond STT-MRAM, we will discuss two approaches: First, we will briefly discuss the giant spin Hall effect as an alternative mechanism to reduce the switching currents of STT-MRAM, by using spin-orbit interactions for inducing large spin currents through magnetic memory bits [48]. More importantly, we will consider magnetoelectric effects that manipulate the magnetization by electric fields (or voltages) instead of current-driven mechanisms [53, 61], allowing for lowest switching energy and increased densities in magnetoelectric RAM (MeRAM) circuits, while keeping the advantages of other types of MRAM (see Table 6.1) [68–70].

6.3 Spin-Transfer Torque (STT) and Its Application in Memory Devices

The discovery of the spin-transfer torque (STT) effect, that is, the possibility to manipulate and induce magnetization switching using spin-polarized currents [16, 17] in nanomagnets, has driven a large number of advances in MRAM in recent years, bringing STT-MRAM [23, 24, 26, 71, 72] closer to practical implementation in electronic products. STT-MRAM offers high speed, very high endurance, and nonvolatility, but the current-driven switching mechanism limits STT-MRAM switching energies to ~100 fJ, much smaller than NAND Flash or magnetic field-switched MRAM, but still much larger than the typical dynamic energy dissipation of CMOS transistors, as will be discussed later.

A simplified STT-MRAM cell structure is shown in Figure 6.1. It consists of a free layer that can be in one of two states, that is, parallel or antiparallel with respect to a pinned (fixed) layer. The layers are separated by a tunneling oxide (usually MgO), which allows for readout via the TMR effect [23, 26]. Writing is performed by passing a current through the device using the access transistor, which induces a magnetization torque that can result in switching depending on the direction of the current.

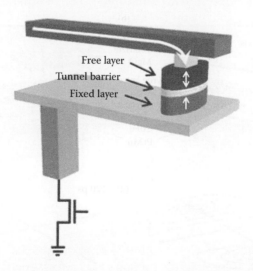

FIGURE 6.1
Typical structure of a 1T-1MTJ STT-MRAM memory cell. The MTJ is composed of a bistable free layer that stores the information, and a pinned layer that is separated from the free layer by a tunneling oxide. The device is accessed by a CMOS transistor as the access device, which also drives currents in order to write information into it. A similar cell structure can be used for MeRAM as well, although VCMA switching also allows for more compact diode-based cells, as will be discussed later. (From K. L. Wang, et al., *Journal of Physics D: Applied Physics*, 46, 074003, 2013.)

Generally one can consider three different configurations of STT-MRAM devices (referred to as I-STT, C-STT, and P-STT, standing for in-plane, combined, and perpendicular, respectively), which offer different performance characteristics. Of particular interest to our current discussion are performance metrics such as speed and switching energy efficiency, which will later be compared to MeRAM devices.

6.3.1 STT-MRAM with In-Plane Free Layers

Figure 6.2 shows two STT-MRAM device configurations where the magnetization of the free layer lies in the plane of the film. The equilibrium states are determined by the shape anisotropy along the easy axis of the nanopillar. In the I-STT configuration, the polarizer (pinned layer) magnetization is in the same plane as the free layer, while in the combined-STT (C-STT) structure (also referred to as orthogonal STT-MRAM, or OST), the material stack includes an additional polarizer that is perpendicular to the sample plane.

In I-STT devices, a key challenge is to reduce the switching current density J_{c0} while keeping a reasonable thermal stability factor Δ for nonvolatility. The value of Δ defines the stability of the magnetic bit against unwanted switching events due to thermal activation (i.e., retention time). As an example, a thermal stability factor of $\Delta = 40$ translates into a mean retention time of about

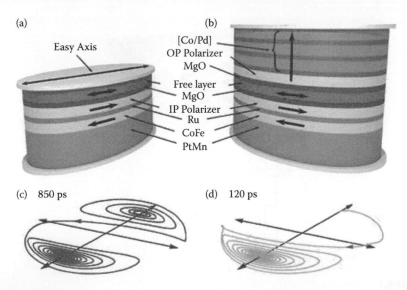

FIGURE 6.2
(a) The layered structure for a typical I-STT device, including a synthetic antiferromagnet (SAF) to compensate for the stray fields from the pinned layer acting on the free layer. (b) The inclusion of an additional perpendicular polarizer results in a C-STT configuration, where the additional polarizer exerts a large initial torque on the free layer magnetization. (c), (d) Macrospin simulations comparing the switching speed of I-STT and C-STT devices. (From G. E. Rowlands, et al., *Applied Physics Letters*, 98, 102509, 2011. With permission.)

10 years for a single device at room temperature. The thermal stability factor is given by $\Delta = E_b/k_B T$, where $E_b = M_s H_k V/2$, M_s is the saturation magnetization, H_k is the magnetic anisotropy field, V is the volume of the free layer, k_B is the Boltzmann constant, and T is the temperature. For in-plane free layers, H_k is mainly determined by the in-plane shape anisotropy. Therefore, as the volume of the free layer is decreased, H_k needs to be increased, for example, by increasing the aspect ratio or thickness of the ellipse, to retain stability. For the I-STT configuration, the switching current density is given by [22, 73] $J_{c0} \approx (2e\alpha M_s t/\hbar\eta)\left(H_k + (H_d - H_{k\perp})/2\right)$, where α is the free layer damping factor, η is the spin transfer efficiency, M_s and t are the free layer saturation magnetization and thickness, H_k is the in-plane shape-induced anisotropy field, $H_d \approx 4\pi M_s \gg H_k$ is the out-of-plane demagnetizing field, and $H_{k\perp}$ is the perpendicular anisotropy field (if any) of the free layer. Scaling of I-STT devices generally requires a trade-off between the critical switching current density J_{c0} and the thermal stability factor Δ, where the goal is to minimize the ratio J_{c0}/Δ while preserving a given thermal stability factor. As will be outlined in later sections, an analogous trade-off exists for MeRAM devices as well.

I-STT suffers from a small spin torque during the initial stages of magnetization switching, a consequence of its parallel equilibrium states. As a result, its switching speed is generally limited to ~1 ns [41, 74]. An alternative to eliminate the initial incubation delay of I-STT devices is to include an additional out-of-plane polarizer (see Figure 6.2(b) and (d)) [41, 43, 74]. This C-STT configuration uses a combination of in-plane and perpendicular polarizers, where the in-plane polarizer provides TMR for readout, while spin torque from the perpendicular polarizer causes the free layer magnetization to undergo a precessional switching process [75–77], allowing for ultrafast switching limited only by the ferromagnetic resonance (FMR) frequency of the free layer. This approach has been shown to allow for switching using current pulses down to ~50 to 100 ps [43, 75, 77, 78]. (In later sections it will be shown how similarly fast precessional switching can be achieved using the VCMA effect, allowing for ultralow power dissipation due to the lower currents involved in that case.) Considering that the write energy is given by $E = V^2 t/R$, the reduction in the required pulse time t can outpace the potential increase in the required voltage for C-STT, resulting in an overall reduction in the switching energy compared to an I-STT device with an otherwise similar layer structure [77].

To further reduce the write energy in STT-MRAM, one can perform further optimization on both the free layer and the tunnel barrier. If the switching energy is written as $E = V^2 t/R = J_C^2 A^2 Rt$, it is evident that it can be further decreased primarily by decreasing the switching current density. Reducing the MgO thickness, and hence the device resistance, also reduces the write energy, but eventually hits a limit imposed by the dielectric barrier quality and its associated breakdown and endurance characteristics [27, 79]. The strategy to decrease the switching energy by reducing the switching current of I-STT devices may be best quantified in terms of the ratio of the switching current to the thermal stability factor, which is given by $I_{c0}/\Delta \approx (4e\alpha k_B T/\hbar\eta H_k)(H_k + (H_d - H_{k\perp})/2)$. Note that the switching current is dominated by the out-of-plane demagnetizing field H_d, which normally, however, does not determine the thermal stability, given that $H_k \ll H_d$. Hence, a promising method to decrease the switching current without sacrificing the thermal stability is to decrease the demagnetizing field H_d by the introduction of a perpendicular anisotropy $H_{k\perp}$ in the free layer [23, 41, 80].

As an example of a promising approach to introduce such a perpendicular anisotropy, which is also particularly relevant to our later discussion of MeRAM devices, significant interface-induced perpendicular magnetic anisotropy (PMA) has been observed and used in a number of material systems in recent years [41, 44]. In particular, this allows for using the familiar CoFeB/MgO material system, which offers large TMR values, while introducing PMA by adjusting the material composition, deposition, and annealing conditions. Using $Co_{20}Fe_{60}B_{20}$ free layers, for example, has been demonstrated to result in a reduction of the average quasi-static switching current density by >40% (from ~2.8 MA/cm^2 to ~1.6 MA/cm^2) due to the presence of the perpendicular anisotropy [41]. Figure 6.3 shows the dependence of the switching current density (at 10 ns) measured for such an MTJ device, on the free layer thickness (which in turn determines the interfacial perpendicular anisotropy), where the fast decrease in switching current with thickness fits well with the interfacial origin of the anisotropy [41].

6.3.2 STT-MRAM with Perpendicular Free Layers

If the perpendicular anisotropies of the free and fixed layers are large enough to overcome their respective demagnetizing fields (i.e., if $H_{k\perp} > H_d$), their respective magnetizations become perpendicular to the sample plane,

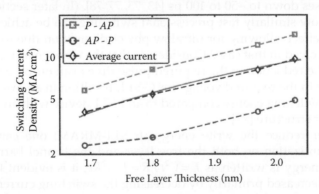

FIGURE 6.3
The interfacial PMA in Fe-rich CoFeB/MgO systems can be exploited to reduce the switching current in I-STT devices. As the free layer thickness is decreased, the perpendicular anisotropy increases due to the interfacial origin of the anisotropy, thereby quickly reducing the switching current. (From P. K. Amiri, et al., *Applied Physics Letters*, 98, 112507, 2011. With permission.)

giving rise to a fully perpendicular (P-STT) configuration [44, 81]. Because the thermal stability factor in P-STT devices is determined by the perpendicular anisotropy of the free layer (and not by its shape as in I-STT), P-STT bits can be patterned into a circular shape [24, 82]. In the P-STT configuration, the switching current density is given by $J_{c0} \approx 2e\alpha M_s (H_{k\perp} - H_d)t/\hbar\eta$, while the figure of merit of switching current to thermal stability ratio is $I_{c0}/\Delta \approx 4e\alpha k_B T/\hbar\eta$. Note that the elimination of the demagnetizing field term in this case leads to a potentially smaller switching current to the thermal stability ratio, if damping can be controlled. Several works have demonstrated P-STT structures utilizing the interfacial perpendicular anisotropy of Fe-rich CoFeB layers in MgO MTJs [28, 44, 83–86], with switching currents as low as <2 MA/cm².

6.3.3 STT-MRAM with Spin Hall-Based MTJ Devices

Despite the potential advantages of STT-MRAM, it still faces a number of fundamental and engineering challenges. In particular, the scaling of MTJ dimensions will be limited by the superparamagnetic limit for in-plane bits, while the development of materials with stronger perpendicular anisotropies is required to maintain the stability in smaller perpendicular devices. However, as discussed in the previous subsection, for P-STT bits the ratio of switching current to thermal stability is determined by fundamental constants and parameters with a limited tuning range (i.e., damping constant and spin transfer efficiency). Therefore, assuming that a constant thermal stability factor (i.e., retention time) will be needed across technology nodes, the switching current will remain nearly constant, independent of the MTJ size, and hence memory cell size will be largely limited by the width of the transistor required to drive the current. (It should be mentioned that this simple single-domain picture is not necessarily correct for larger MTJs, where a reduction in switching current with scaling has been observed with a constant thermal stability factor. It is, however, reliable at scaled nodes below ~30 nm.)

In order to provide a better scaling scenario, as well as to reduce the dynamic energy dissipation, the possibility of exploiting spin-orbit interactions, such as in the giant spin Hall effect, has been suggested to switch the magnetization in MTJs [48]. It has recently been demonstrated that a current flowing through a metallic layer with large spin-orbit coupling (SOC), such as Ta or W adjacent to a CoFeB free layer, can generate a spin current through the MTJ, which is large enough to induce switching in the device [48]. The device structure is shown in Figure 6.4, where the electrical current J_C flowing in the high-SOC conductor induces a spin current $J_S = \theta_{SH} J_C$ through the memory bit, where θ_{SH} is referred to as the spin Hall angle. Note that the direction of the current J_C controls the spin polarization of J_S, and hence the switching direction will be determined by the direction of J_C. By

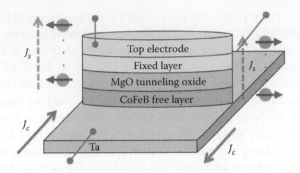

FIGURE 6.4
Memory device configuration utilizing the spin Hall effect. An advantage of this configuration is that the required switching current density can be engineered by tuning the lateral area of the wire, while the bit area can be designed to account for a given thermal stability factor. (From K. L. Wang, et al., *Journal of Physics D: Applied Physics*, 46, 074003, 2013.)

using high-SOC materials such as W and Ta, spin Hall angles θ_{SH} as large as ~30% have been realized without further increase in the damping of the adjacent CoFeB free layer [47, 48].

Due to its three-terminal structure, an SHE memory device potentially results in a penalty in terms of density compared to a two-terminal MTJ. However, it also offers a number of advantages. First, the switching current density required to switch the MTJ can be engineered by the dimensions of the high-SOC metal wire without disturbing the thermal stability of the MTJ [47, 48, 87]. Second, the read and write paths are separated in this structure, potentially allowing disturb-free read operations. Finally, in terms of switching energy, the smaller switching current flowing through a low-resistance metallic wire can lead to switching energy levels smaller than in regular STT-MRAM. However, the need for ultrathin metal interconnects can limit the overall size of memory arrays in this approach, as well as their energy efficiency, given the high resistance associated with them. While the use of spin-orbit torques for magnetization switching remains a thriving area of research, we will next turn our attention to voltage-controlled magnetic devices, which are the main focus of this chapter.

6.4 Voltage-Controlled Magnetic Anisotropy (VCMA) Effect in Magnetic Films

It is known that interfaces of magnetic thin films with nonmagnetic materials exhibit magnetoelectric (ME) effects; that is, their magnetic properties are sensitive to electric fields at the interface [57–59, 88]. In particular, interfaces of oxides with metallic ferromagnets such as Fe, Co, or CoFeB can exhibit a large interfacial perpendicular magnetic anisotropy (PMA) [28, 41, 44, 89, 90], which is also sensitive to voltages applied across the dielectric layer [58, 59, 61, 62, 64, 88, 91–98]. This effect has been attributed to the electric field-induced modulation of the relative occupancy of different orbitals at the interface, which, combined with the spin-orbit coupling, results in a change of the magnetic anisotropy energy [58, 59, 61, 88]. If MgO is used as the dielectric barrier, this voltage-controlled anisotropy effect can be incorporated into MTJ devices with high TMR ratios [4, 5, 95, 96, 99–101] for applications such as memory. However, as will be discussed later, there have also been attempts at integrating dielectric materials with larger dielectric constants, such as GdO_x and TaO_x into VCMA structures, to increase the effect of the applied voltage on interface charge accumulation, and hence magnetic anisotropy [102–105].

The PMA effect in a $MgO/Co_{20}Fe_{60}B_{20}/Ta$ material structure is illustrated in Figure 6.5, where a transition of the film magnetization from in-plane to out-of-plane is observed as the $Co_{20}Fe_{60}B_{20}$ layer thickness is reduced [41]. The inset in Figure 6.5(b) shows the measured total perpendicular anisotropy as a function of the film thickness, with a reorientation at ~1.5 nm. Above this transition thickness, the shape-induced demagnetizing field (favoring in-plane magnetization) is larger than the interfacial anisotropy (which favors perpendicular magnetization). This interfacial PMA effect has been used to realize fully perpendicular magnetic tunnel junctions [44, 84, 86, 106], to reduce switching current in in-plane MTJs [41], and to increase output power in spin torque nano-oscillators [30, 33]. Near the transition thickness, however, the magnetization configuration is also sensitive to voltages applied to the device [95, 107–110], and can therefore be used to realize electric field-controlled magnetic memory devices.

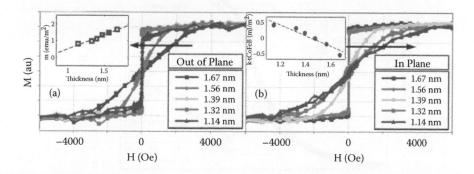

FIGURE 6.5
See color insert. Magnetization as a function of magnetic field, measured with (a) out-of-plane and (b) in-plane magnetic fields on films with varying CoFeB thickness. Inset in (a) shows the measured magnetic moment as a function of film thickness. Inset in (b) shows the dependence of perpendicular anisotropy on thickness, consistent with an interfacial origin of the perpendicular anisotropy. (From P. K. Amiri, et al., *Applied Physics Letters*, 98, 112507, 2011. With permission.)

The interfacial VCMA effect has been investigated using a number of methods. One approach to characterize its magnitude is using magnetometry measurements on a magnetic film as a function of applied electric field bias [61], or (equivalently) to measure magnetoresistance curves of the device under different bias conditions [91]. This approach has been used in [91], for example, where values for the VCMA-induced anisotropy field modulation of ~600 Oe/V, equivalent to a change in anisotropy energy of 37 fJ/V-m, have been obtained. Figure 6.6 illustrates the corresponding experiment as described in [91]. The fixed layer is in-plane in this experiment, while the free layer thickness is tuned such that its magnetization is out-of-plane, resulting in a hard-axis behavior in the magnetoresistance loop when in-plane magnetic fields are applied. The slope of the curves in Figure 6.6 can be used to infer the perpendicular anisotropy [91], and is a function of the voltage bias applied to the device (compare, for example, curves for +80 and −80 mV bias).

In [92], a similar anisotropy modulation by electric fields is measured using the anomalous Hall effect. A $Co_{40}Fe_{40}B_{20}$ free layer composition is used in this case, without the presence of a fixed layer, and both cases of in-plane and out-of-plane magnetization are studied. Magneto-optical Kerr effect (MOKE) measurements were used in [94] to study the VCMA effect in $Co_{20}Fe_{80}$ films, also in the absence of a pinned layer. Most experimental results for CoFe/ MgO and similar material systems to date have reported VCMA magnitudes of ~30 to 40 fJ/V-m, in general agreement with theoretical predictions [98].

As will be outlined in later sections, for device applications a large VCMA parameter is of critical interest, as it translates into a larger change in magnetic

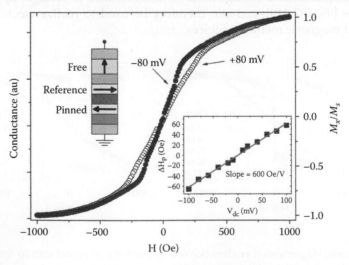

FIGURE 6.6
Voltage-induced modulation of the perpendicular anisotropy (VCMA) in the free layer of a magnetic tunnel junction. The inset shows the dependence of the perpendicular anisotropy on applied voltage. (From J. Zhu, et al., *Physical Review Letters*, 108, 197203, 2012. With permission.)

anisotropy with a lower voltage, resulting in, for example, smaller switching voltages for magnetic memory. To this end, a number of works have investigated the effect of various oxides, magnetic free layer materials, and capping or seed layers on the VCMA effect. In MTJ devices with an Fe-rich CoFeB free layer, the voltage-controlled interfacial effect can generate effective fields as large as 600 Oe per volt [91]. A magnetoelectric coefficient close to the one observed in CoFeB/MgO systems has also been observed experimentally for the FePt/MgO system [111]. Enhancement of the magnetoelectric coefficient has been demonstrated by stacking MgO next to a material with a larger dielectric constant [112, 113]. The observed enhancement is proportional to the increase in the effective dielectric constant of the stack, as expected (e.g., ~30% in $MgO/HfO_2/Al_2O_3$ stacks [112]). Further, a very large enhancement of the magnetoelectric coefficient, compared to CoFeB/MgO systems, has been obtained recently in experiments by using FePd next to MgO [60], where VCMA parameters as large as 600 fJ/V-m have been measured. It has also been shown that capping or seed layers placed next to the ferromagnet play a critical role in determining VCMA. As an example, VCMA larger than 1100 fJ/V-m has been measured in epitaxial V/Fe/MgO films [114]. Finally, modulation of anisotropy by electric fields has also been observed in GdO_x [115], AlO_x, and TaO_x [97], the latter with a magnetoelectric coefficient fairly similar to CoFeB/MgO, despite its larger dielectric constant.

In many cases, the exact reasons for the observed VCMA values, and their dependence on parameters such as the seed layer, film compositions, or oxide layers, are not fully understood at the moment and require further investigation. In each case there may be several different factors at play, including spin-orbit coupling of the materials used, interdiffusion of layers affecting the ferromagnetic film composition, formation of interfacial oxides, as well as possible ionic transport through the oxide in addition to the electron accumulation/depletion at the interface, all of which can affect the magnetic properties. The challenge of finding a ferromagnetic material (FM)/oxide system with enhanced voltage-controlled magnetic anisotropy (VCMA) effects and simultaneously large magnetoresistance is currently an important topic of research. It is a complex problem with multiple facets requiring both significant experimental effort and insight from first-principles calculations, to identify material systems with reliably large VCMA effects.

6.5 Voltage-Induced Ferromagnetic Resonance Excitation

The interfacial VCMA effect has also been studied by microwave FMR measurements [62, 64, 91, 93]. Early experiments demonstrated the modulation of high-frequency magnetization dynamics using DC voltages, where the voltage led to an observable shift in the FMR [62] and spin wave [93] frequencies, yet the high-frequency dynamics themselves were generated by means other than electric fields (e.g., by spin-transfer torque).

Similar to the case of switching, the use of currents (rather than voltages using VCMA) to induce high-frequency magnetization dynamics is energetically inefficient and leads to larger ohmic losses. This is a disadvantage if radio frequency (RF) dynamics need to be excited with high energy efficiency, such as in RF device applications [29, 35, 36, 116, 117], as well as in potential applications in spin wave logic circuits [118, 119]. For these practical reasons as well as for better fundamental understanding of the VCMA effect, the study of the microwave electric field-induced dynamics in magnetic nanostructures is of great interest. Several works have addressed this, including in $Co_{20}Fe_{60}B_{20}$-MgO [91] and $Co_{20}Fe_{80}$-MgO [64] magnetic tunnel junctions, where RF magnetization dynamics were directly generated through the application of an RF voltage to the device using the VCMA effect.

The experiment described in [91] is illustrated in Figure 6.7. The $Co_{20}Fe_{60}B_{20}$ free layer of the device is thin enough to point out of the sample plane, while the reference (pinned) layer is in-plane. The FMR signal is detected by measuring the rectified voltage across the device, when a microwave voltage is applied to it, in an approach similar to [62, 64, 116, 120]. Due to the relatively low resistance of the MTJs used in this particular experiment, both VCMA and STT effects were present. They were characterized separately by analysis of the resonance line shape [91]. Based on this, STT- and VCMA-induced

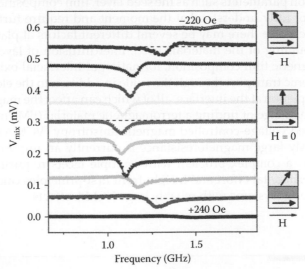

FIGURE 6.7
VCMA-induced high-frequency magnetization dynamics in a magnetic tunnel junction. FMR spectra are obtained by measuring the rectified voltage across the device when a high-frequency voltage is applied to it. Both STT and VCMA effects are present. Their relative magnitudes can be identified using the resonance line shape. (From J. Zhu, et al., *Physical Review Letters*, 108, 197203, 2012. With permission.)

torques were determined to be approximately equal in magnitude in this low-resistance device, and the magnitude of the VCMA effect at frequencies of a few gigaHertz was comparable to that at DC, hence affecting the high-frequency (or short-timescale) behavior of the device. With increased device resistance, the relative contribution of the VCMA effect to the device behavior becomes larger compared to the current-induced STT effect.

Results such as those described in [64, 91] demonstrate that the VCMA effect can be used directly to generate RF dynamics in high-resistance MTJs, where STT is suppressed due to small leakage currents, potentially conferring an advantage in terms of energy efficiency. As an example, [91] shows that the sensitivity of an MTJ microwave signal detector with combined VCMA and STT effects present can be increased by ~40%, reaching values close to those of Schottky diodes. Equally importantly, however, these results indicate the significant role played by VCMA even in regular current-controlled STT devices, such as low-resistance MTJ memory cells used for STT-MRAM. This is especially the case in STT-MRAM devices with large interfacial PMA, where VCMA is often present as well [41, 44].

More specifically, however, the VCMA effect itself can also be used as the primary physical mechanism to bring about switching in magnetoelectric MRAM (i.e., MeRAM) memory cells that do not utilize the STT effect, which have resistances high enough to prevent the flow of large currents, with advantages in terms of energy efficiency and density (due to smaller access transistor size) compared to STT-MRAM. We will occasionally refer to such VCMA-dominated MTJs as VMTJs or MEJs (for magnetoelectric tunnel junction) in this text, to distinguish them from STT-dominated MTJs, although depending on the context, the term *MTJ* can still be applied. The next section reviews some of the experiments on VCMA-induced switching of such MEJs, aimed at the realization of a voltage-controlled nonvolatile magnetic memory.

6.6 Voltage-Induced Switching of Magnetic Memory Bits

Several reports have demonstrated the feasibility of using VCMA to induce or assist in the switching of magnetic memory devices [70, 87, 95, 96, 101, 107, 108, 121]. Compared to STT-induced switching, VCMA-induced switching offers the potential to reduce power dissipation, and enhance density and scalability by eliminating the need for large drive currents (hence large access devices). It also offers a pathway to applications beyond memory where superior energy efficiency is required. Figure 6.8 (top) illustrates a VCMA-controlled magnetic memory bit, with free and fixed layers as in a regular MTJ, allowing for the electrical readout of the memory via the TMR effect. The leakage current through the device is small, allowing for the applied voltage (electric field across the barrier) to control the device behavior.

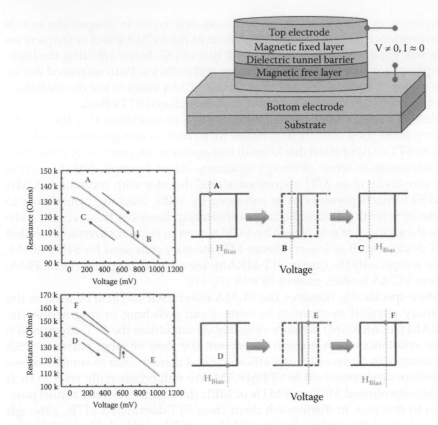

FIGURE 6.8
See color insert. Schematic illustration of a voltage-controlled magnetic tunnel junction (top) using the VCMA effect [87, 107, 108]. Bottom panel shows the voltage-induced switching of a nanoscale device with in-plane magnetization. The switching process is assisted by a small magnetic field that also determines the switching direction. Note that the same voltage polarity is used for switching in both directions.

The voltage control of the perpendicular anisotropy can be used to induce switching in MEJ memory devices in both thermal and precessional regimes. When ultrafast (<1 ns) voltage pulses are applied, the voltage-induced magnetization dynamics can be used to generate precessional switching by proper timing of the pulse, analogous to the C-STT switching discussed previously. In a second approach, the VCMA effect can be used to modulate the coercivity of the MEJ devices, allowing for electric field-assisted switching. As illustrated in Figure 6.8, in this case the applied voltage induces a reduction in the free layer coercivity due to the modification of the interface anisotropy, translating into a single available state for a given applied bias magnetic field. Note that the direction of the switching therefore will be determined by the effective applied field, while the switching is obtained for the same voltage polarity in both directions. The role of the external field can be replaced by allowing a nonzero leakage current to pass though the device [108].

Equivalent to the coercivity-based picture, the switching mechanism in such a VCMA-based device can be described as follows (assuming an in-plane magnetization in both free and fixed layers, while the case of a fully perpendicular device follows similarly). Switching is realized through the modification of the PMA when a voltage pulse is applied, that is, by turning the free layer from a stable in-plane state to an intermediate perpendicular state for a normally in-plane magnetized bit. Once the voltage pulse is removed, the free layer magnetization relaxes to one of the available in-plane states, with the final state being determined by the total magnetic field acting on the free layer. This total field can be a combination of stray fields from the fixed layer, as well as any external fields present in the experiment.

An example of switching data for such a voltage-assisted switching experiment are shown in Figure 6.8 [96, 107, 108]. While two different external magnetic fields (+60 Oe and −60 Oe) are used to bring about switching in different directions in this in-plane device, the switching itself is induced by VCMA, rather than STT. This is evident from the large device resistance (corresponding to leakage currents of <10 µA during switching). It is also worth noting that unlike STT, both switching directions are realized using voltages of the same polarity in this experiment, further pointing to VCMA as the primary switching mechanism [107, 108]. A similar switching experiment based on this thermally activated principle for fully perpendicular MTJs is described in [96]. Here, both magnetic electrodes consist of $Co_{40}Fe_{40}B_{20}$, and can be switched separately using voltages of opposite polarities.

The VCMA effect can also be used to bring about precessional switching of the magnetization in magnetic tunnel junctions. This is qualitatively similar to current-induced precessional switching realized in STT-MRAM devices [41, 43, 74, 78], with very fast switching times in the subnanosecond range, but with much better energy efficiency due to the electric field-controlled nature of switching. This type of switching is, however, sensitive to the precise width of the applied pulse [95, 101], with the final state being determined by the number of precessions undergone by the free layer magnetization while the voltage is applied (see Figure 6.9). A similar type of switching for perpendicular devices is described in [101].

VCMA effects have also been incorporated into different types of devices, which may or may not be described as MRAM cells, depending on the detailed implementation. One example is the voltage control of domain wall velocity in ultrathin magnetic films [97], with potential applications to domain wall logic [122] and racetrack memory [123]. Another example is a magnetoelectric charge trap memory [102], where VCMA is used to provide a magnetic readout for a charge trapping-based nonvolatile memory device. The device contains a ZrO_2 charge trap layer, where holes from an adjacent electrode are trapped upon application of a voltage. Once the voltage is removed, the trapped charge creates an electric field across a dielectric

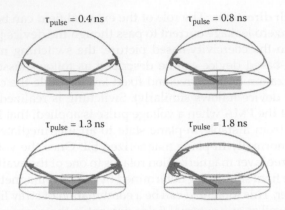

FIGURE 6.9
Electric field-induced precessional switching of magnetization in a magnetic tunnel junction.
Figure shows macrospin simulations demonstrating the dependence of the final state on the
applied pulse width. (From Y. Shiota, et al., *Nature Materials*, 11, 39–43, 2012. With permission.)

MgO layer, affecting the perpendicular anisotropy of a thin Fe film beneath
it. The device operates similar to a charge trap Flash memory, except for the
readout, which is magnetic and performed via the interfacial VCMA effect.

6.7 Analysis of VCMA-Induced Switching

The VCMA effect can be utilized to switch magnetic tunnel junctions as
described in the previous section. In this section, we present an analysis
of this electric field-induced switching in the thermally activated regime,
following closely the discussion presented in [69]. Figure 6.10 shows a
schematic representation of MEJs with out-of-plane and in-plane magnetizations in the free and fixed layers. Switching is accomplished by putting
the free layer in an intermediate state using an applied voltage pulse. This
is realized through the modification of perpendicular anisotropy when
the voltage is applied, for example, turning the free layer from a stable
in-plane state to a perpendicular state for a normally in-plane magnetized
bit. When the voltage is removed, the free layer will relax to one of the
stable in-plane states, which is determined by the overall magnetic field
acting on the free layer. This field may be a combination of coupling fields
provided by the fixed layer (e.g., through dipole coupling), as well as any
external fields present. The case of a perpendicularly magnetized device
follows similarly. In the following, we will consider the critical voltage
required for switching using this VCMA-based approach, assuming a single-domain nanomagnet.

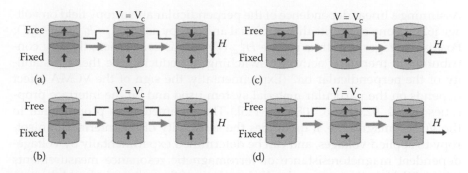

FIGURE 6.10
Schematic illustration of magnetoelectric tunnel junctions switched by the combination of an applied voltage and external magnetic field, with out-of-plane (a, b) and in-plane (c, d) magnetization in the free and fixed layers [69]. The final state depends on the direction of the overall magnetic field acting on the free layer, and can be controlled by applying an external field. The role of the external field can be replaced by current-induced torques, for example, by allowing for a small leakage current to pass through the device.

6.7.1 Perpendicular Bits

Consider the case of a memory bit with perpendicular (out-of-plane) magnetization in both free and fixed layers. The thermal stability factor of such a memory bit is given by

$$\Delta = E_b/k_B T \tag{6.1}$$

where k_B is the Boltzmann constant and T is temperature. Similar to the case of other magnetic memories [22, 23, 26], Δ is a measure of the retention time (hence nonvolatility) of the device. E_b is the energy barrier between the two free layer states, which is given by

$$E_b = H_{k,eff}^\perp(V)M_s At/2 \tag{6.2}$$

where M_s is the free layer saturation magnetization, A is its area, and t is its thickness. The effective perpendicular anisotropy field is given by

$$H_{k,eff}^\perp(V) = h_{k,s}^\perp(V)/t - 4\pi M_s \tag{6.3}$$

where $h_{k,s}^\perp(V)/t$ is the thickness-dependent interfacial perpendicular magnetic anisotropy field [41, 44], which can be modulated by the applied voltage V [61, 91, 92, 98]. Note that Equation (6.3) assumes the perpendicular anisotropy to be entirely interfacial in origin, as is the case in many material systems of interest [28, 41, 44, 91]. The energy barrier is a function of voltage, and the standby thermal stability factor is thus determined by the value of $H_{k,eff}^\perp(0)$.

Assuming a linear dependence of the perpendicular anisotropy field on volt-age [62, 91], one can write the interfacial anisotropy as $h_{k,s}^{\perp}(V) = h_{k,s}^{\perp}(0)(1-\zeta V)$. Positive voltages will thus reduce $H_{k,eff}^{\perp}(V)$ in this convention, thereby con-tributing to thermally activated switching by reducing the thermal stabil-ity of the perpendicular bit.[*] (Experimentally, the sign of the VCMA effect depends on the particular material system used and on the interface prop-erties [61, 91, 92, 95, 96, 108, 124, 125].) The parameter ζ is proportional to the VCMA magnitude, it quantifies the sensitivity of the interfacial anisot-ropy to applied voltages, and can be determined experimentally by voltage-dependent magnetoresistance or ferromagnetic resonance measurements [61, 64, 91]. Note that a more detailed analysis of the voltage-induced dynam-ics should also include the second-order magnetic anisotropy, especially in cases where the first-order perpendicular anisotropy field $h_{k,s}^{\perp}(0)/t$ and the demagnetizing field $4\pi M_s$ are nearly equal in Equation (6.3) [91]. We neglect this point in the present work, assuming thereby that the standby condition characterized by $H_{k,eff}^{\perp}(0)$ is sufficiently far from the cancellation point of the first-order anisotropy and the demagnetizing field.

The voltage-dependent thermal stability factor of the magnetic bit is thus given by

$$\Delta(V) = \Delta(0) - M_s A \left(\zeta V h_{k,s}^{\perp}(0) + tH \right) / 2 k_B T \tag{6.4}$$

where

$$\Delta(0) = \left(h_{k,s}^{\perp}(0) - 4\pi M_s t \right) M_s A / 2 k_B T \tag{6.5}$$

is the standby thermal stability of the bit, and where we have accounted for a bias field H applied externally to the bit, as in the case of [96, 108]. Note that the direction (i.e., sign) of H determines the switching direction, by increas-ing or decreasing $\Delta(V)$ depending on the initial state of the magnetic bit. The voltage-induced modulation of the perpendicular magnetic anisotropy hence results in a voltage-dependent dwell time for the magnetic bit, allow-ing for thermally activated switching on a timescale of

$$\tau(V) = \tau_0 \exp(\Delta(V)) \tag{6.6}$$

where τ_0 is the attempt time. Hence, the thermally activated switching time reduces as the voltage is increased. Note that voltages of the opposite polar-ity will increase $\Delta(V)$, hence further stabilizing the bit. The critical switching voltage V_c, which corresponds to $\tau(V_c) = \tau_0$, can then be obtained from the condition $\Delta(V_c) = 0$, and is given by

---

[*] Note: however, the validity of the present analysis does not depend on the sign of the VCMA effect.

$$V_c = \left(h_{k,s}^{\perp}(0) - (4\pi M_s + H)t\right)/\zeta h_{k,s}^{\perp}(0) \tag{6.7}$$

Note that the contribution of external magnetic fields is thus not only to determine the switching direction (through the sign of H), but also to affect the critical switching voltage. In the limit where the VCMA effect is negligible ($\zeta \approx 0$), Equation (6.4) thus reduces to a critical switching field of $H_c = H_{k,eff}^{\perp}(0)$, as expected.

The role of external fields can thus be incorporated into this model by accounting for the magnetic field H, which acts to reduce or increase the energy barrier, depending on its direction with respect to the initial state of the device, while the voltage-dependent VCMA effect acts to reduce or increase the barrier depending only on the voltage polarity. This is illustrated in Figure 6.11.

6.7.2 In-Plane Bits

The case of a voltage-controlled nanomagnet with in-plane magnetizations in both free and fixed layers can be treated similarly. In this case, we assume an elliptical monodomain magnetic bit, where the energy barrier E_b (and hence the thermal stability factor) is determined by the shape-induced anisotropy field H_{shape} and is given by

$$E_b = H_{shape} M_s A t / 2 \tag{6.8}$$

in which

$$H_{shape} \approx 8\pi M_s t \beta / w \tag{6.9}$$

where w is the width of the elliptical bit, and $\beta = (r-1)/r$ is a shape factor determined by the in-plane aspect ratio r [82].

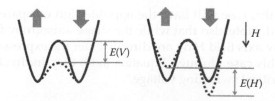

FIGURE 6.11
Schematic representation of the effect of applied voltages (left) and external magnetic fields (right) on the energy diagram of a bistable magnetic bit. The external magnetic field reduces or increases the energy barrier, depending on its orientation with respect to the initial state of the magnetization in the bit, while VCMA reduces or increases it for both switching directions, depending only on the voltage polarity [69]. The reduced energy barrier results in a modification of the dwell time, resulting in thermally activated switching. Within this picture, precessional switching is achieved for voltages large enough to eliminate the energy barrier, hence moving the magnetization from its initial state and setting it on a precessional trajectory.

In this case, because the applied voltage still only modifies the out-of-plane anisotropy of the bit, the thermally activated switching time is only affected by the applied voltage if the perpendicular anisotropy is sufficiently increased, so that it provides a lower energy barrier for switching compared to that given by Equation (6.8), such that

$$4\pi M_s - h_{k,s}^{\perp}(V)/t < H_{shape} + H \tag{6.10}$$

Note that in this case, the magnetic bias field H is oriented in-plane and along the easy axis of the ellipse, and for typical experimental conditions, $H \ll H_{shape}$. For voltages that satisfy Equation (6.10), the thermal stability factor thus will become a function of the applied voltage, and will be given by

$$\Delta(V) = \left(4\pi M_s t - h_{k,s}^{\perp}(0)(1 - \zeta V)\right) M_s A / 2 k_B T \tag{6.11}$$

where

$$4\pi M_s - h_{k,s}^{\perp}(0)/t > H_{shape} \tag{6.12}$$

in order to ensure standby stability of the bit. The voltage bringing about switching thus has the opposite polarity compared to that of a perpendicularly magnetized bit (e.g., it is a negative voltage for a positive ζ for the same sign of VCMA), such as to increase the perpendicular anisotropy and to bring about switching. The voltage-induced modulation of the thermal stability thus results in a voltage-dependent dwell time for the magnetic bit, similar to Equation (6.6). The critical switching voltage V_c, which corresponds to $\tau(V_c) = \tau_0$, is thus given by

$$V_c = \left(h_{k,s}^{\perp}(0) - 4\pi M_s t\right) / \zeta h_{k,s}^{\perp}(0) \tag{6.13}$$

Note that V_c in this case will have the opposite sign compared to Equation (6.7), as expected. Note also that while the shape anisotropy field H_{shape} and the in-plane easy-axis field H do not directly enter the expression for switching voltage in this case, through Equation (6.10) they indirectly set a lower bound for the critical switching voltage.

6.7.3 Discussion and Comparison to STT Switching

Figure 6.12 shows the critical switching voltage V_c as a function of the perpendicular interface anisotropy $h_{k,s}^{\perp}(0)$ for a perpendicular magnetic bit, as given by Equation (6.7). The results are based on a device with $t = 1$ nm, $4\pi M_s = 1$ T, the out-of-plane field H ranging from 100 to 500 Oe, and experimentally accessible [91] values for ζ ranging from 0.06 to 0.12 V^{-1}. It can be seen that the externally applied magnetic field H contributes to reducing the switching voltage.

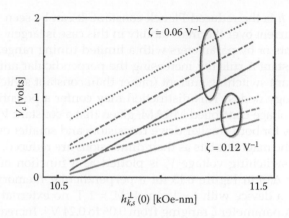

FIGURE 6.12
The switching voltage as a function of perpendicular interface anisotropy for a perpendicular magnetic bit. Device parameters are $t = 1$ nm, $4\pi M_s = 1$ T, $H = 100$ Oe (dotted), 300 Oe (dashed), and 500 Oe (solid). Two values for ζ are compared. The case of $\zeta = 0.06$ V^{-1} corresponds to a voltage-induced modulation of the anisotropy energy by ~36 fJ/V-m for a film with $t = 1.5$ nm and $4\pi M_s = 1$ T, consistent with present experimental values. (From P. K. Amiri, et al., *Journal of Applied Physics*, 113, 013912, 2013. With permission.)

Increasing the VCMA effect (i.e., larger ζ) also reduces the switching voltage as expected, and furthermore reduces the sensitivity of the switching voltage to the applied external field.

An important parameter that determines the scaling behavior of nonvolatile magnetic memories is the ratio of switching voltage (or current) over the thermal stability factor, which should ideally be minimized. For perpendicularly magnetized voltage-controlled memory bits, Equation (6.7) can be rewritten as $V_c = \left(\left(2k_B T \Delta(0)/M_s A \right) - tH \right) / \zeta h_{k,s}^\perp(0)$, which for $H \approx 0$ results in

$$\frac{V_c}{\Delta(0)} = \frac{2k_B T}{\zeta M_s h_{k,s}^\perp(0) A} \qquad (6.14)$$

As with other magnetic memories, scaling to smaller bit areas A requires a corresponding increase in the perpendicular anisotropy $h_{k,s}^\perp(0)$ to maintain a constant thermal stability $\Delta(0)$ (hence nonvolatile retention time) of the bits. (It is also possible to maintain a constant $\Delta(0)$ by using materials with a larger bulk contribution of perpendicular anisotropy, but this reduces the VCMA parameter ζ, thereby increasing V_c and resulting in a less favorable scaling behavior.) Provided that $h_{k,s}^\perp(0)$ can be sufficiently increased, it is in principle possible to maintain an approximately constant value for $V_c/\Delta(0)$ across technology nodes for MEJs. It is interesting to compare this to the corresponding scaling parameter for perpendicular STT-based memory, which as mentioned before, is given by $I_c/\Delta(0) = 4e\alpha k_B T/\hbar\eta$, where I_c is the switching current, e is the electron charge, α is the Gilbert damping constant, η is the spin transfer efficiency, and $\hbar$ is the reduced Planck constant. It can be seen that the ratio of switching current over thermal stability in this case is largely set by fundamental constants or by parameters with a limited tuning range. Hence, scaling with a constant Δ rule (by increasing the perpendicular anisotropy) will lead to a constant switching current (rather than constant switching voltage) across technology nodes. This distinction may confer a potential scalability advantage to voltage-controlled MRAM, given that a constant $V_c/\Delta(0)$ scaling scenario allows for both smaller drive transistors and smaller overall switching energy of the magnetic bits as their dimensions are reduced.

The critical switching voltage V_c is plotted as a function of the thermal stability factor $\Delta(0)$ in Figure 6.13 for a perpendicular memory bit. Results are shown for a device with $t = 1$ nm, $4\pi M_s = 1$ T, no external field ($H = 0$), and the VCMA parameter ζ ranging from 0.06 to 0.24 V^{-1}. Increasing ζ results in a smaller V_c for a given thermal stability, thus pushing the curve further toward the practically relevant area on the lower right corner of the plot (e.g., $\Delta(0) > 40$ and $V_c < 1$ V).

From Equations (6.6), (6.7), and (6.11) one can obtain an expression for the switching voltage as a function of the mean switching time, given by

$$V = V_c\left(1 - \Delta(0)^{-1}\ln(\tau/\tau_0)\right) \qquad (6.15)$$

One can compare this model to experimental switching results obtained from nanoscale devices. The results are given in Figure 6.14, which shows measured switching results for antiparallel (AP) to parallel (P) and P to AP switching in a 60×170 nm bit with a resistance-area (RA) product of ~1,300 $\Omega\text{-}\mu m^2$, that is, greater than 100 times larger than in typical STT

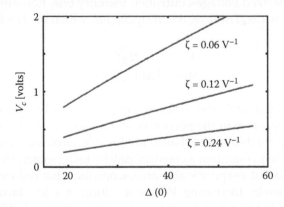

FIGURE 6.13
The switching voltage as a function of the thermal stability factor for a perpendicular magnetic bit. Device parameters are $t = 1$ nm, $4\pi M_s = 1$ T, and $H = 0$. Three values for ζ are compared, indicating an improving trade-off between thermal stability and switching voltage with increasing VCMA. (From P. K. Amiri, et al., *Journal of Applied Physics*, 113, 013912, 2013. With permission.)

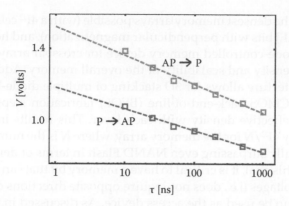

FIGURE 6.14
Measured switching voltages for pulse widths down to 10 ns, for a 60×170 nm voltage-controlled magnetoelectric tunnel junction. Dashed lines represent the fits to Equation (6.15), giving $V_{c,P-AP} = 1.18$ V and $V_{c,AP-P} = 1.53$ V. (From P. K. Amiri, et al., *Journal of Applied Physics*, 113, 013912, 2013. With permission.)

devices [23, 26, 27]. Leakage currents in the experiment were limited to <10 μA, minimizing the effect of STT and allowing for switching dynamics dominated by VCMA [108]. Different switching directions were realized using external fields ($H = +80$ Oe and $H = -80$ Oe for different switching directions, respectively). Figure 6.14 shows a good fit to the thermal activation model of Equation (6.15) for both switching directions, with critical switching voltages of $V_{c,P-AP} = 1.18$ V and $V_{c,AP-P} = 1.53$ V.

6.8 Circuit Implementations of VCMA-Based MeRAM

In this section we will examine circuit implications of electric field-controlled MeRAM and will focus in particular on diode-based array implementations, following closely the discussion presented in [70]. STT-MRAM normally uses a one-transistor/one-MTJ (1T-1MTJ) cell structure with CMOS transistors as the access devices (see Figure 6.1). However, the relatively large currents required to switch STT-based MTJs require large transistors to drive them [70]. As a result, the density of STT-MRAM arrays is often limited not by the dimensions of their MTJs, but rather by the switching current of the magnetic bit itself. Additionally, the use of three-terminal CMOS transistors also imposes a layout-based limit of approximately $8F^2$ on the maximum cell density. The use of transistors in STT-MRAM is dictated by its purely current-controlled write mechanism, where currents of opposite polarities are needed to write different bits of information, preventing the realization of crossbar arrays with diodes as the access devices. In principle, however, crossbars are the densest memory arrays possible (with a $4F^2$ cell size, assuming circular MEJ bits with perpendicular magnetization), and hence the realization of a diode-controlled memory device for crossbar arrays can greatly increase the density and scalability of the overall memory. Additionally, the crossbar architecture allows for 3D stacking of multiple diode-MEJ memory layers in the CMOS back-end-of-line (BEOL) fabrication steps, potentially doubling the effective density with each layer. This results in an effective cell size of only $4F^2/N$ for the memory array, where N is the number of layers, hence potentially surpassing even NAND Flash in terms of density.

To achieve this goal, it is critical to have a memory bit that can be controlled by unipolar voltages (i.e., does not require opposite directions of current), so that a diode can be used as the access device. As discussed in previous sections, VCMA offers this possibility. In particular, some designs proposed for MeRAM realize this requirement by using MEJs with a voltage-controlled magnetic anisotropy (VCMA) switching mechanism, combined with STT [70, 108]. The use of VCMA allows the MEJ to switch in both directions (write opposite bits of information) using voltages of the same polarity, but with different magnitudes. This enables a diode-MEJ crossbar architecture, with the potential for a sub-$1F^2$ effective cell size.

Figure 6.15 shows how voltage pulses of the same polarity, but different amplitudes, can be used to switch the device in opposite directions [70, 108]. No external magnetic field is required. A positive voltage pulse above a certain threshold forces the free layer into a metastable intermediate state. When this voltage is removed, the free layer will relax to either the parallel (P) or antiparallel (AP) state, depending on the amplitude of the applied voltage and its resulting STT current. While the switching is voltage induced (i.e., by VCMA), the final state is determined by the combination of STT and

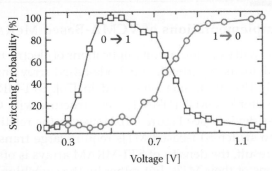

FIGURE 6.15
Measured probability of switching curves for VCMA-based MEJs with STT-assisted switching. The combination of VCMA and STT effects allows for a unipolar set/reset switching scheme with switching voltages of 0.5 and 1.1 V, respectively. (From R. Dorrance, et al., *IEEE Electron Device Letters*, 34, 753–755, 2013. With permission. J. G. Alzate, et al., Voltage-Induced Switching of Nanoscale Magnetic Tunnel Junctions, presented at the *IEEE International Electron Devices Meeting (IEDM)*, San Francisco, CA, 2012. With permission.)

the coupling field from the pinned layer (which are designed to oppose each other). Hence a small, nonzero leakage current passes through the device, which, while not sufficient to bring about purely STT-induced switching by itself, contributes to determining the switching direction. This allows for a unipolar set/reset write scheme, where voltage pulses of the same polarity, but different amplitudes, can be used to switch the device between the P and AP states. Voltage pulses of the opposite polarity will not switch the device, but rather reinforce the initial state [70, 108]. The probability of writing as a function of applied voltage for the MEJs used in a prototype crossbar array is shown in Figure 6.15, demonstrating their unipolar characteristics. The devices measure 190 × 60 nm, corresponding to a 65-nm CMOS technology.

Figure 6.16 shows the schematic and layout view for one vertical slice of a high-density crossbar memory array using voltage-controlled MEJs. The unipolar set/reset write scheme of the devices allows for a diode to be integrated in series without any loss in functionality. The series diode also has the added benefit of eliminating the sneak currents present in traditional crossbar arrays. These parasitic currents ultimately limit the performance and maximum size of the memory array. By placing the memory arrays directly over the array decoders and sense amplifiers, a cell efficiency of nearly 100% can potentially be achieved.

During both reading and writing, unaccessed bit-lines (BLs) are grounded while unaccessed source-lines (SLs) are pulled to *VDD* (1.5 V), reverse biasing the series diode for unaccessed bits. During the write operation, the target SL is pulled to ground, while the target BL is pulsed with the appropriate set/reset voltage, that is, 0.5 and 1.1 V, respectively. During the read operation, the target SL is pulled to ground and the target BL is connected

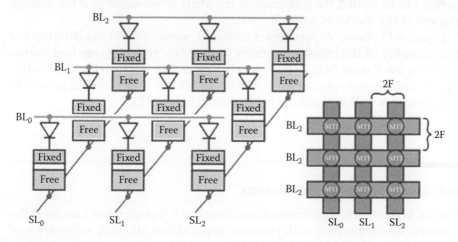

FIGURE 6.16
Schematic and layout views of a crossbar array structure, with integrated diodes, for high-density 3D MeRAM. (From R. Dorrance, et al., *IEEE Electron Device Letters*, 34, 753–755, 2013. With permission.)

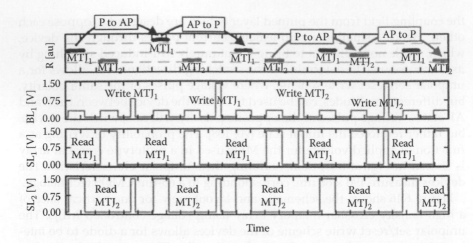

FIGURE 6.17

Measured transient waveforms for reading and writing of addressed memory cells in a MeRAM crossbar array, demonstrating that two different cells can be written selectively with unipolar set/reset voltages (0.5 and 1.1 V, respectively), and read without disturbing adjacent bits. (From R. Dorrance, et al., *IEEE Electron Device Letters*, 34, 753–755, 2013. With permission.)

to a sense amplifier. To prevent disturbing the state of the desired bit cell, a sensing voltage of 0.2 V is used. In order to maximize the read margin of the accessed memory cell, minimizing the forward drop across the access diode is crucial. This can be accomplished through the use of low-threshold Schottky diodes. The ultimate size of the crossbar array is determined by the reverse bias leakage of the diodes and the resistance of the MEJs. Larger arrays can be built if the resistance of the MEJs is increased or if the leakage current of the diodes is reduced.

Figure 6.17 shows experimental transient waveforms demonstrating the functionality of the crossbar memory array. MEJ_1 and MEJ_2 are first initialized into the P state. MEJ_1 is first switched from P to AP, then back to P, without disturbing the value of MEJ_2. Similarly, MEJ_2 is also switched from P to AP, then back to P, without disturbing the value of MEJ_1.

6.9 Summary and Conclusions

The VCMA effect has generated a significant and fast-growing interest in the magnetism community, with potential applications affecting many areas of spintronics research, and in particular magnetic memory.

In the area of memory, VCMA may complement or replace STT as the primary switching mechanism used in MRAM, due to several reasons. First, STT-MRAM applications may eventually be limited by the power dissipation associated with this current-controlled mechanism, which, while sufficiently low to be of interest for random access memory applications, may limit the usefulness of STT-based devices for applications beyond memory, such as nonvolatile logic. Second, the typical transistor-based cell structures used for STT-MRAM, along with the currents required to switch STT-MRAM bits, impose a limitation on its scalability and density. VCMA-based devices could offer solutions to both of these issues, as outlined in this chapter. By allowing for both thermally activated and precessional switching, VCMA can access a wide range of access times for different applications, and can be used to realize ultrafast write times in magnetic memory cells down to the subnanosecond regime. Successful realization of these potential advantages, however, will require development of improved material structures with larger VCMA effects (i.e., operating at lower voltages), while maintaining other important attributes, such as high TMR. Finally, by providing a fundamentally new mechanism for coupling the electrical and magnetic domains in nanomagnetic structures, VCMA will allow for novel spintronic device concepts as well as spark new directions for fundamental research in magnetism.

References

1. S. A. Wolf, et al., Spintronics: a spin-based electronics vision for the future, *Science*, 294, 1488–1495, 2001.
2. S. Ikeda, et al., Magnetic tunnel junctions for spintronic memories and beyond, *IEEE Transactions on Electron Devices*, 54, 991–1002, 2007.
3. M. Julliere, Tunneling between ferromagnetic films, *Physics Letters A*, 54, 225–226, 1975.
4. S. S. P. Parkin, et al., Giant tunnelling magnetoresistance at room temperature with MgO (100) tunnel barriers, *Nature Materials*, 3, 862–867, 2004.
5. S. Yuasa, et al., Giant room-temperature magnetoresistance in single-crystal Fe/MgO/Fe magnetic tunnel junctions, *Nature Materials*, 3, 868–871, 2004.
6. W. H. Butler, et al., Spin-dependent tunneling conductance of Fe|MgO|Fe sandwiches, *Physical Review B*, 63, 054416, 2001.
7. J. Mathon and A. Umerski, Theory of tunneling magnetoresistance of an epitaxial Fe/MgO/Fe(001) junction, *Physical Review B*, 63, 220403, 2001.
8. G. Binasch, et al., Enhanced magnetoresistance in layered magnetic structures with antiferromagnetic interlayer exchange, *Physical Review B*, 39, 4828–4830, 1989.
9. M. N. Baibich, et al., Giant magnetoresistance of (001)Fe/(001)Cr magnetic superlattices, *Physical Review Letters*, 61, 2472–2475, 1988.
10. P. Khalili Amiri and K. L. Wang, Voltage-controlled magnetic anisotropy in spintronic devices, *Spin*, 2(3), 2012.
11. M. Tondra, et al., Picotesla field sensor design using spin-dependent tunneling devices, *Journal of Applied Physics*, 83, 6688–6690, 1998.
12. X. Liu, et al., Magnetic tunnel junction field sensors with hard-axis bias field, *Journal of Applied Physics*, 92, 4722–4725, 2002.

13. D. Lacour, et al., Field sensing using the magnetoresistance of IrMn exchange-biased tunnel junctions, *Journal of Applied Physics*, 91, 4655–4658, 2002.

14. Z. M. Zeng, et al., Nanoscale magnetic tunnel junction sensors with perpendicular anisotropy sensing layer, *Applied Physics Letters*, 101, 062412, 2012.

15. P. Wisniowski, et al., Effect of free layer thickness and shape anisotropy on the transfer curves of MgO magnetic tunnel junctions, *Journal of Applied Physics*, 103, 07A910, 2008.

16. J. C. Slonczewski, Current-driven excitation of magnetic multilayers, *Journal of Magnetism and Magnetic Materials*, 159, L1–L7, 1996.

17. L. Berger, Emission of spin waves by a magnetic multilayer traversed by a current, *Physical Review B*, 54, 9353–9358, 1996.

18. J. A. Katine, et al., Current-driven magnetization reversal and spin-wave excitations in Co/Cu/Co pillars, *Physical Review Letters*, 84, 3149–3152, 2000.

19. Y. M. Huai, et al., Observation of spin-transfer switching in deep submicron-sized and low-resistance magnetic tunnel junctions, *Applied Physics Letters*, 84, 3118–3120, 2004.

20. M. Tsoi, et al., Generation and detection of phase-coherent current-driven magnons in magnetic multilayers, *Nature*, 406, 46–48, 2000.

21. K. L. Wang and P. Khalili Amiri, Nonvolatile spintronics: Perspectives on instant-on nonvolatile nanoelectronic systems, *SPIN*, 2(2), 2012.

22. J. A. Katine and E. E. Fullerton, Device implications of spin-transfer torques, *Journal of Magnetism and Magnetic Materials*, 320, 1217–1226, 2008.

23. E. Chen, et al., Advances and future prospects of spin-transfer torque random access memory, *IEEE Transactions on Magnetics*, 46, 1873–1878, 2010.

24. Y. Huai, Spin-transfer torque MRAM (STT-MRAM): Challenges and prospects, *AAPPS Bulletin*, 18, 33–40, 2008.

25. C. J. Lin, et al., 45nm low power CMOS logic compatible embedded STT MRAM utilizing a reverse-connection 1T/1MTJ cell, in *2009 IEEE International Electron Devices Meeting (IEDM)*, 2009, pp. 1–4.

26. K. Lee and S. H. Kang, Development of embedded STT-MRAM for mobile system-on-chips, *IEEE Transactions on Magnetics*, 47, 131–136, 2011.

27. P. K. Amiri, et al., Low write-energy magnetic tunnel junctions for high-speed spin-transfer-torque MRAM, *IEEE Electron Device Letters*, 32, 57–59, 2011.

28. D. C. Worledge, et al., Spin torque switching of perpendicular Ta | CoFeB | MgO-based magnetic tunnel junctions, *Applied Physics Letters*, 98, 022501, 2011.

29. Q. Mistral, et al., Current-driven microwave oscillations in current perpendicular-to-plane spin-valve nanopillars, *Applied Physics Letters*, 88, 192507, 2006.

30. Z. Zeng, et al., High-power coherent microwave emission from magnetic tunnel junction nano-oscillators with perpendicular anisotropy, *ACS Nano*, 2012.

31. S. Kaka, et al., Mutual phase-locking of microwave spin torque nano-oscillators, *Nature*, 437, 389–392, 2005.

32. D. Houssameddine, et al., Spin-torque oscillator using a perpendicular polarizer and a planar free layer, *Nature Materials*, 6, 447–453, 2007.

33. Z. M. Zeng, et al., Ultralow-current-density and bias-field-free spin-transfer nano-oscillator, *Scientific Reports*, 3, No. 1426, 2013.

34. V. S. Pribiag, et al., Magnetic vortex oscillator driven by d.c. spin-polarized current, *Nature Physics* 3, 498–503, 2007.

35. S. I. Kiselev, et al., Microwave oscillations of a nanomagnet driven by a spin-polarized current, *Nature*, 425, 380–383, 2003.

36. W. H. Rippard, et al., Direct-current induced dynamics in Co_{90}Fe_{10}/Ni_{80}Fe_{20} point contacts, *Physical Review Letters*, 92, 027201, 2004.

37. F. B. Mancoff, et al., Phase-locking in double-point-contact spin-transfer devices, *Nature*, 437, 393–395, 2005.

38. F. B. Ren and D. Markovic, True energy-performance analysis of the MTJ-based logic-in-memory architecture (1-bit full adder), *IEEE Transactions on Electron Devices*, 57, 1023–1028, 2010.

39. S. Matsunaga, et al., Standby-power-free compact ternary content-address-able memory cell chip using magnetic tunnel junction devices, *Applied Physics Express*, 2, 023004, 2009.

40. S. Matsunaga, et al., Fabrication of a nonvolatile full adder based on logic-in-memory architecture using magnetic tunnel junctions, *Applied Physics Express*, 1, 091301, 2008.

41. P. K. Amiri, et al., Switching current reduction using perpendicular anisotropy in CoFeB-MgO magnetic tunnel junctions, *Applied Physics Letters*, 98, 112507, 2011.

42. H. Zhao, et al., Low writing energy and sub nanosecond spin torque transfer switching of in-plane magnetic tunnel junction for spin torque transfer random access memory, *Journal of Applied Physics*, 109, 070720, 2011.

43. H. Liu, et al., Ultrafast switching in magnetic tunnel junction based orthogonal spin transfer devices, *Applied Physics Letters*, 97, 242510, 2010.

44. S. Ikeda, et al., A perpendicular-anisotropy CoFeB-MgO magnetic tunnel junction, *Nature Materials*, 9, 721–724, 2010.

45. S. Y. Lee, et al., A full adder design using serially connected single-layer magnetic tunnel junction elements, *IEEE Transactions on Electron Devices*, 55, 890–895, 2008.

46. H. Meng, et al., A spintronics full adder for magnetic CPU, *IEEE Electron Device Letters*, 26, 360–362, 2005.

47. C.-F. Pai, et al., Spin transfer torque devices utilizing the giant spin Hall effect of tungsten, *Applied Physics Letters*, 101, 122404, 2012.

48. L. Liu, et al., Spin-torque switching with the giant spin Hall effect of tantalum, *Science*, 336, 555–558, 2012.

49. W. Eerenstein, et al., Multiferroic and magnetoelectric materials, *Nature*, 442, 759–765, 2006.

50. R. Ramesh and N. A. Spaldin, Multiferroics: Progress and prospects in thin films, *Nature Materials*, 6, 21–29, 2007.

51. M. Gajek, et al., Tunnel junctions with multiferroic barriers, *Nature Materials*, 6, 296–302, 2007.

52. G. Srinivasan, et al., Magnetoelectric bilayer and multilayer structures of mag-netostrictive and piezoelectric oxides, *Physical Review B*, 64, 214408, 2001.

53. T. Wu, et al., Electrical control of reversible and permanent magnetization reori-entation for magnetoelectric memory devices, *Applied Physics Letters*, 98, 262504, 2011.

54. H. Ohno, et al., Electric-field control of ferromagnetism, *Nature*, 408, 944–946, 2000.

55. D. Chiba, et al., Electrical control of the ferromagnetic phase transition in cobalt at room temperature, *Nature Materials*, 10, 853–856, 2011.

56. I. V. Ovchinnikov and K. L. Wang, Voltage sensitivity of Curie temperature in ultrathin metallic films, *Physical Review B*, 80, 012405, 2009.

57. M. K. Niranjan, et al., Electric field effect on magnetization at the Fe/MgO(001) interface, *Applied Physics Letters*, 96, 222504, 2010.

58. C.-G. Duan, et al., Surface magnetoelectric effect in ferromagnetic metal films, *Physical Review Letters*, 101, 137201, 2008.
59. J. P. Velev, et al., Multi-ferroic and magnetoelectric materials and interfaces, *Philosophical Transactions of the Royal Society A: Mathematical, Physical and Engineering Sciences*, 369, 3069–3097, 2011.
60. F. Bonell, et al., Large change in perpendicular magnetic anisotropy induced by an electric field in FePd ultrathin films, *Applied Physics Letters*, 98, 232510, 2011.
61. T. Maruyama, et al., Large voltage-induced magnetic anisotropy change in a few atomic layers of iron, *Nature Nanotechnology*, 4, 158–161, 2009.
62. T. Nozaki, et al., Voltage-induced perpendicular magnetic anisotropy change in magnetic tunnel junctions, *Applied Physics Letters*, 96, 022506, 2010.
63. J. Zhu, et al., Voltage-induced ferromagnetic resonance in magnetic tunnel junctions, *Physical Review Letters*, 108, 197203, 2012.
64. T. Nozaki, et al., Electric-field-induced ferromagnetic resonance excitation in an ultrathin ferromagnetic metal layer, *Nature Physics*, 8, 492–497, 2012.
65. M. Mayberry, *Emerging technologies and Moore's law: prospects for the future*, 2010, csg5.nist.gov/pml/div683/upload/Mayberry_March_2010.pdf.
66. G. E. Moore, Cramming more components onto integrated circuits, *Electronics* magazine, 38(8), 1965.
67. S. Assefa, et al., Fabrication and characterization of MgO-based magnetic tunnel junctions for spin momentum transfer switching, *Journal of Applied Physics*, 102, 063901, 2007.
68. P. Khalili Amiri and K. L. Wang, Voltage-controlled magnetic anisotropy in spintronic devices, *SPIN*, 2, 1240002, 2012.
69. P. K. Amiri, et al., Electric-field-induced thermally assisted switching of monodomain magnetic bits, *Journal of Applied Physics*, 113, 013912, 2013.
70. R. Dorrance, et al., Diode-MTJ crossbar memory cell using voltage-induced unipolar switching for high-density MRAM, *IEEE Electron Device Letters*, 34, 753–755, 2013.
71. *International Technology Roadmap for Semiconductors*, 2005, ed.
72. K. L. Wang, et al., From nanoelectronics to nano-spintronics, *Journal of Nanoscience and Nanotechnology*, 11, 306–313, 2011.
73. J. Z. Sun, Spin-current interaction with a monodomain magnetic body: A model study, *Physical Review B*, 62, 570–578, 2000.
74. A. D. Kent, et al., Spin-transfer-induced precessional magnetization reversal, *Applied Physics Letters*, 84, 3897–3899, 2004.
75. D. Bedau, et al., Ultrafast spin-transfer switching in spin valve nanopillars with perpendicular anisotropy, *Applied Physics Letters*, 96, 022514, 2010.
76. D. Bedau, et al., Spin-transfer pulse switching: From the dynamic to the thermally activated regime, *Applied Physics Letters*, 97, 262502, 2010.
77. G. E. Rowlands, et al., Deep subnanosecond spin torque switching in magnetic tunnel junctions with combined in-plane and perpendicular polarizers, *Applied Physics Letters*, 98, 102509, 2011.
78. O. J. Lee, et al., Spin-torque-driven ballistic precessional switching with 50 ps impulses, *Applied Physics Letters*, 99, 102507, 2011.
79. Z. M. Zeng, et al., Effect of resistance-area product on spin-transfer switching in MgO-based magnetic tunnel junction memory cells, *Applied Physics Letters*, 98, 072512, 2011.

80. L. Q. Liu, et al., Reduction of the spin-torque critical current by partially cancel-ing the free layer demagnetization field, *Applied Physics Letters*, 94, 122508, 2009.
81. H. Meng and J. P. Wang, Spin transfer in nanomagnetic devices with perpen-dicular anisotropy, *Applied Physics Letters*, 88, 172506, 2006.
82. D. Apalkov, et al., Comparison of scaling of in-plane and perpendicular spin transfer switching technologies by micromagnetic simulation, *IEEE Transactions on Magnetics*, 46, 2240–2243, 2010.
83. M. T. Rahman, et al., Reduction of switching current density in perpendicu-lar magnetic tunnel junctions by tuning the anisotropy of the CoFeB free layer, *Journal of Applied Physics*, 111, 07C907, 2012.
84. D. C. Worledge, et al., Spin torque switching of perpendicular Ta | CoFeB | MgO-based magnetic tunnel junctions, *Applied Physics Letters*, 98, 022501, 2011.
85. M. T. Rahman, et al., High temperature annealing stability of magnetic proper-ties in MgO-based perpendicular magnetic tunnel junction stacks with CoFeB polarizing layer, *Journal of Applied Physics*, 109, 07C709, 2011.
86. M. T. Rahman, et al., Reduction of switching current density in perpendicu-lar magnetic tunnel junctions by tuning the anisotropy of the CoFeB free layer, *Journal of Applied Physics*, 111, 07C907, 2012.
87. K. L. Wang, et al., Low-power non-volatile spintronic memory: STT-RAM and beyond, *Journal of Physics D: Applied Physics*, 46, 074003, 2013.
88. C. G. Duan, et al., Tailoring magnetic anisotropy at the ferromagnetic/ferroelec-tric interface, *Applied Physics Letters*, 92, 122905, 2008.
89. S. Yakata, et al., Influence of perpendicular magnetic anisotropy on spin-trans-fer switching current in CoFeB/MgO/CoFeB magnetic tunnel junctions, *Journal of Applied Physics*, 105, 07D131, 2009.
90. W.-G. Wang, et al., Rapid thermal annealing study of magnetoresistance and perpendicular anisotropy in magnetic tunnel junctions based on MgO and CoFeB, *Applied Physics Letters*, 99, 102502, 2011.
91. J. Zhu, et al., Voltage-induced ferromagnetic resonance in magnetic tunnel junc-tions, *Physical Review Letters*, 108, 197203, 2012.
92. M. Endo, et al., Electric-field effects on thickness dependent magnetic anisot-ropy of sputtered MgO/Co(40)Fe(40)B(20)/Ta structures, *Applied Physics Letters*, 96, 212503, 2010.
93. S. S. Ha, et al., Voltage induced magnetic anisotropy change in ultrathin Fe(80) Co(20)/MgO junctions with Brillouin light scattering, *Applied Physics Letters*, 96, 142512, 2010.
94. Y. Shiota, et al., Voltage-assisted magnetization switching in ultrathin Fe(80) Co(20) alloy layers, *Applied Physics Express*, 2, 063001, 2009.
95. Y. Shiota, et al., Induction of coherent magnetization switching in a few atomic layers of FeCo using voltage pulses, *Nature Materials*, 11, 39–43, 2012.
96. W.-G. Wang, et al., Electric-field-assisted switching in magnetic tunnel junc-tions, *Nature Materials*, 11, 64–68, 2012.
97. A. J. Schellekens, et al., Electric-field control of domain wall motion in perpen-dicularly magnetized materials, *Nature Communications*, 3, 847, 2012.
98. M. K. Niranjan, et al., Electric field effect on magnetization at the Fe/MgO(001) interface, *Applied Physics Letters*, 96, 222504, 2010.
99. Y. Shiota, et al., Pulse voltage-induced dynamic magnetization switching in magnetic tunneling junctions with high resistance-area product, *Applied Physics Letters*, 101, 102406, 2012.

100. J. G. Alzate, et al., Voltage-induced switching of nanoscale magnetic tunnel junctions, *IEDM Technical Digest*, 29.5.1–29.5.4, 681–684, December, 2012.
101. S. Kanai, et al., Electric field-induced magnetization reversal in a perpendicular-anisotropy CoFeB-MgO magnetic tunnel junction, *Applied Physics Letters*, 101, 122403, 2012.
102. U. Bauer, et al., Magnetoelectric charge trap memory, *Nano Letters*, 12, 1437–1442, 2012.
103. U. Bauer, et al., Electric field control of domain wall propagation in Pt/Co/GdOx films, *Applied Physics Letters*, 100, 192408, 2012.
104. U. Bauer, et al., Voltage-controlled domain wall traps in ferromagnetic nanowires, *Nature Nanotechnology*, 8, 411–416, 2013.
105. A. J. Schellekens, et al., Electric-field control of domain wall motion in perpendicularly magnetized materials, *Nature Communications*, 3, No. 847, 2012.
106. M. Gajek, et al., Spin torque switching of 20 nm magnetic tunnel junctions with perpendicular anisotropy, *Applied Physics Letters*, 100, 132408, 2012.
107. J. G. Alzate, et al., Voltage-induced switching of CoFeB-MgO magnetic tunnel junctions, in *56th Conference on Magnetism and Magnetic Materials*, Scottsdale, AZ, 2011, p. EG-11.
108. J. G. Alzate, et al., Voltage-induced switching of nanoscale magnetic tunnel junctions, presented at the *IEEE International Electron Devices Meeting (IEDM)*, San Francisco, CA, 2012.
109. F. Bonell, et al., Large change in perpendicular magnetic anisotropy induced by an electric field in FePd ultrathin films, *Applied Physics Letters*, 98, 232510, 2011.
110. W. G. Wang, et al., Electric-field-assisted switching in magnetic tunnel junctions, *Nature Materials*, 11, 64–68, 2012.
111. T. Seki, et al., Coercivity change in an FePt thin layer in a Hall device by voltage application, *Applied Physics Letters*, 98, 212505, 2011.
112. K. Kita, et al., Electric-field-control of magnetic anisotropy of $Co[_{0.6}]Fe[_{0.2}]B[_{0.2}]$/oxide stacks using reduced voltage, *Journal of Applied Physics*, 112, 033919, 2012.
113. M. Endo, et al., Electric-field effects on thickness dependent magnetic anisotropy of sputtered $MgO/Co[_{40}]Fe[_{40}]B[_{20}]$/Ta structures, *Applied Physics Letters*, 96, 212503, 2010.
114. A. Rajanikanth, et al., Magnetic anisotropy modified by electric field in V/Fe/MgO(001)/Fe epitaxial magnetic tunnel junction, *Applied Physics Letters*, 103, 062402, 2013.
115. U. Bauer, et al., Electric field control of domain wall propagation in Pt/Co/GdOx films, *Applied Physics Letters*, 100, 192408, 2012.
116. A. A. Tulapurkar, et al., Spin-torque diode effect in magnetic tunnel junctions, *Nature*, 438, 339–342, 2005.
117. X. Fan, et al., Magnetic tunnel junction based microwave detector, *Applied Physics Letters*, 95, 122501, 2009.
118. A. Khitun, et al., Magnonic logic circuits, *Journal of Physics D—Applied Physics*, 43, 264005, 2010.
119. A. Khitun and K. L. Wang, Non-volatile magnonic logic circuits engineering, *Journal of Applied Physics*, 110, 034306, 2011.
120. J. C. Sankey, et al., Spin-transfer-driven ferromagnetic resonance of individual nanomagnets, *Physical Review Letters*, 96, 227601, 2006.

121. Y. Shiota, et al., Pulse voltage-induced dynamic magnetization switching in magnetic tunneling junctions with high resistance-area product, *Applied Physics Letters*, 101, 102406, 2012.
122. D. A. Allwood, et al., Magnetic domain-wall logic, *Science*, 309, 1688–1692, 2005.
123. S. S. P. Parkin, et al., Magnetic domain-wall racetrack memory, *Science*, 320, 190–194, 2008.
124. Y. Shiota, et al., Opposite signs of voltage-induced perpendicular magnetic anisotropy change in CoFeB | MgO junctions with different underlayers, *Applied Physics Letters*, 103, 082410, 2013.
125. S. Kanai, et al., Electric field-induced magnetization reversal in a perpendicular-anisotropy CoFeB-MgO magnetic tunnel junction, *Applied Physics Letters*, 101, 122403, 2012.

121. Y. Shiota, et al. Pulse voltage-induced dynamic magnetization switching in magnetic tunneling junctions with high resistance-area product. Applied Physics Letters, 101, 102406, 2012.

122. D. A. Allwood, et al. Magnetic domain-wall logic. Science, 309, 1688–1692, 2005.

123. S. S. P. Parkin, et al. Magnetic domain-wall racetrack memory. Science, 320, 190–194, 2008.

124. Y. Shiota, et al. Opposite sense of voltage-induced perpendicular magnetic anisotropy change in CoFeB|MgO junctions with different underlayers. Applied Physics Letters, 103, 082410, 2013.

125. S. Kanai, et al. Electric field-induced magnetization reversal in a perpendicular anisotropic CoFeB-MgO magnetic tunnel junction. Applied Physics Letters, 101, 122403, 2012.

Index

Printed and bound by CPI Group (UK) Ltd, Croydon, CR0 4YY

18/10/2024

01776208-0007